CONJURING THE UNIVERSE

Peter Atkins is a fellow of Lincoln College in the University of Oxford and the author of about seventy books for students and a general audience. His texts are market leaders around the globe. A frequent lecturer in the United States and throughout the world, he has held visiting professorships in France, Israel, Japan, China, and New Zealand. He was the founding chairman of the Committee on Chemistry Education of the International Union of Pure and Applied Chemistry and was a member of IUPAC's Physical and Biophysical Chemistry Division. He was the 2016 recipient of the American Chemical Society's Grady-Stack Award for the communication of chemistry to the public.

Praise for *Conjuring the Universe*

'Taking on nothing less than the laws of nature, Atkins seeks to answer the question of how they came into being, and in doing so, composes an elegant love letter to the simplicity and beauty of the mathematics that govern our universe... Atkins sweeps aside the mathematical mystique with his characteristic wit. He leads even the most unmathematical of readers (this one included) to appreciate his argument that the universe, in all its breathtaking intricacy, probably emerged from a nothing in which anything that can happen does, and in which our ignorance of the behaviour of individual entities leads us to the basis of everything we understand about our world.'

Zoë Hackett, *Chemistry World*

'It's rare to find a study of physical laws that is also a bravura display of rarefied humour and experiential depth; but such is this gem by chemist Peter Atkins.'

Barbara Kiser, *Nature*

'The discussion of constants of Nature is interesting and presented a bit differently than most such, especially with regard to the question of which constants are truly fundamental. I enjoyed reading the book, not only for the main themes but also for several asides on history, etymology, and so on.'

Philip Helbig, *Observatory Magazine*

'Atkins writes in a charming, even chummy way. He understands our confusion and leads us onwards with the promise of great insights: how the very laws of physics came to be... *Conjuring the Universe* is a clear example of [Atkins's] extraordinary erudition and flair.'

Robyn Williams, *Australian Book Review*

'Atkins writes in a clear and humorous manner for the lay reader. Don't skip the notes at the end of the book. Some real gems are hidden there... Recommended for undergraduates and general readers.'

Choice

'Tour de force... this is a compact 168 pages that delivers splendidly on the question of where the natural laws came from.'

Brian Clegg, popularscience.com

CONJURING *the* UNIVERSE

the origins of the laws of nature

PETER ATKINS

OXFORD
UNIVERSITY PRESS

OXFORD
UNIVERSITY PRESS

Great Clarendon Street, Oxford, OX2 6DP,
United Kingdom

Oxford University Press is a department of the University of Oxford.
It furthers the University's objective of excellence in research, scholarship,
and education by publishing worldwide. Oxford is a registered trade mark of
Oxford University Press in the UK and in certain other countries

First published 2018
First published in paperback 2020
Impression: 1

Published in the United States of America by Oxford University Press
198 Madison Avenue, New York, NY 10016, United States of America

British Library Cataloguing in Publication Data
Data available

Library of Congress Cataloging in Publication Data
Data available

ISBN 978–0–19–881337–8 (Hbk.)
ISBN 978–0–19–881338–5 (Pbk.)

Printed and bound in Great Britain by
Clays Ltd, Elcograf S.p.A.

PREFACE

The workings of the world have been ascribed by some to an astonishingly busybody but disembodied Creator, actively guiding every electron, quark, and photon into its destiny. My gut recoils from this profligate vision of the workings of the world, and my head follows where my gut points. In the following pages I explore whether there is a simpler vision of what is going on. Scientists, after all, are hewers of simplicity from complexity, and typically prefer the less elaborate to the more. I explore the furthest recesses exposed by this hewing, and argue that the laws of nature, our summaries of the workings of the world, arose in the simplest possible way. I argue that they spring from nothing more than indolence and anarchy, spiced up here and there by a dash of ignorance.

I have limited my arena of discourse to the commonplace. Thus I take you through mechanics, both classical and quantum, thermodynamics, and electromagnetism. With diminishing confidence, but I hope in a thought-provoking manner, I lead you into the origin of the fundamental constants and conclude by wondering about the efficacy of mathematics in the formulation of natural laws and its possible exposure of the deep structure of reality. Some of what I write is speculation, for full understanding is still beyond science's grasp despite the remarkable progress it has made in three centuries of serious endeavour. If you seek a deeper justification of some of my remarks, then you will find it in the

Notes at the end of the book, which I have treated as a safe space for harbouring equations.

I hope in the following pages that I answer questions that might have occurred to you but have gone unanswered, and that my words will reveal some aspects of the astonishing simplicity of this gloriously complex world.

PETER ATKINS
Oxford, 2017

CONTENTS

For all knowledge and wonder (which is the seed of knowledge) is an impression of pleasure in itself.

—Francis Bacon, *The Advancement of Learning* (1605)

1

Back to Eternity

The nature of laws

I need to prepare your mind for a thought so absurd it might even be true. Science has progressed through revolutions, sometimes called 'paradigm shifts', when what had been taken to be common sense or was simply a prevailing attitude has been displaced by something seemingly closer to the truth. Aristotle provides one example, Copernicus another. Thus, Aristotle from his marble armchair reflected on the flight of arrows and inferred that they were pushed along by vortices in the air behind them. He also took the more general view, by noticing how carts were kept in motion by the constant exercise of effort by oxen, that motion had to be sustained by effort. Galileo and then Newton saw through the effects of atmosphere and mud and replaced Aristotle's vision by its opposite, in which motion continues effortlessly unless quenched by effort. The air inhibits the flight of arrows, and although Aristotle could not know it, arrows fly even better in a vacuum where there can be no sustaining vortices. He ought to have noticed, but did not have the opportunity, that carts on ice instead of getting bogged down in mud do not need the pull of oxen to sustain their motion. Copernicus, as is well known, effected a cosmic revolution, and a profound simplification of understanding (a significant marker of nearing the truth), when he rejected the common-sense daily revolution of the Sun about the Earth and settled on a central, stationary Sun and an orbiting, spinning Earth.

More subtle but no less far-reaching revisions were to come with the intellectual revolutions of the early twentieth century dwarfing the nearly contemporary political upheavals and those of a century or so earlier. The common-sense view that events can be regarded as simultaneous had to be abandoned after 1905 once Albert Einstein (1879–1955) had transformed our perception of space and time, blending them together into spacetime. That blending stirred time into space and space into time to an extent that depended on the speed of the observer. With time and space so entangled, no two observers in relative motion could agree on whether two events were simultaneous. This fundamental revision of the arena of our actions and perceptions might seem to be a heavy price to pay for approaching greater understanding, but it too turns out to yield a simplification of the mathematical description of the physical world: no longer did explanations of phenomena need to be cobbled together from the *bricolage* of the concepts of Newtonian physics, they emerged naturally from the melding of space with time.

At about the same time, the newly hatched class of quantum theorists transformed thought in another direction by showing that Newton had been deceived in yet another way and that even Einstein's migration of Newtonian physics into his new arena of spacetime was fundamentally false. In that case, although he had scraped the effect of mud off movement, Newton had been trapped in the common-sense, farmyard-inspired vision that to specify a path it was necessary to consider both position and velocity. How aghast were the classical physicists, even those who had learned to live contentedly in spacetime, when it turned out that this notion had to be discarded. In the popular mind the icon of that discarding is the uncertainty principle, formulated by Werner Heisenberg (1901–76) in 1927. The principle asserts that position and velocity cannot be known simultaneously, and so seemed to undermine all hope of understanding by eliminating what had been taken to be the underpinning of nature, or at least the underpinning of the description of nature. Later in the book I shall argue against the view that Heisenberg's principle undermines the prospect of understanding and complete description.

There was apparently worse to come (but like many conceptual upsets, that worse was actually better in worse's clothing). Common sense distinguished unhesitatingly between particles and waves. Particles were little knobbly jobs; waves waved. But in a revolution that shook matter to its core the distinction was found to be false. An early example was the discovery in 1897 of the electron by the physicist J. J. Thomson (1856–1940) with all the attributes of a particle, but then, in 1911, the demonstration by his son G. P. Thomson (1892–1975), among others, that on the contrary the electron had all the attributes of a wave. I like to imagine the father and son sitting icily silent at their breakfast.

More evidence accumulated. A ray of light, without doubt a wave of electromagnetic radiation, was found to share the attributes of a stream of particles. Particles undulated into waves according to the type of observation made on them; waves likewise corpusculated into particles. Once quantum mechanics had become established (in 1927), largely by Heisenberg monkishly isolated on an island and Erwin Schrödinger (1887–1961), as he reported, in an outpouring of erotic passion while on a mountain with a mistress, no longer could this fundamental distinction be preserved:

in a totally non-commonsensical way all entities—from electrons upwards—had a dual, blended nature. Duality had usurped identity.[1]

I could go on. The more the deep structure of the world is exposed, the less it seems that common sense—by which I mean intuition based on local, uncontrolled, casual experience of the everyday environment, essentially the undigested food of the gut rather than the intellectual food of assessing collective brains, in other words the controlled, detailed inspection of isolated fragments of the world (in short, experiments)—is a reliable source of information. More and more it seems that deeper understanding comes from sloughing off each layer of common sense (but of course, retaining rationality). With that in mind, and your mind I hope prepared to relinquish common sense as an investment in attaining future comprehension, I would like to displace one further aspect of common sense.

I would like to assert that not much happened at the Creation. I am aware, of course, of the compelling descriptions that have endowed this moment with awesome drama, for surely the birth of everything ought to have been cosmically dramatic? A giant cosmic cataclysm of an event. A spectacular universe-wide burst of amazing primordial activity. A terrific explosion rocking spacetime to its foundations. A pregnant fireball of searing, space-burning intensity. Really, really big. The 'Big Bang', the name itself, evokes drama on a cosmic scale. Indeed, Fred Hoyle (1915–2001) introduced the term dismissively and sardonically in 1949, favouring his own theory of ongoing, continuous, serenely perpetual creation, eternal cosmogenesis, world without beginning entailing world without end. The Bang is viewed as a massive explosion filling all space, in fact creating all space and all time, and in the tumult of heat, the entire universe expanding from a mere dot of unimaginable temperature and density into the far cooler and huge, still-expanding extent we regard as our cosmic home today. Add to that the current vogue for considering an 'inflationary era', when the universe doubled in size every tiny fraction of a second before reaching, after less than a blink of an eye, the middle-aged relatively demure expansion, with temperatures of only a few million degrees, that began the era we know today.

Not much happening? Yes, it is a big step to think of all that hyperactivity, energy, and emergence of fundamental stuff in general as representing not much happening. But bear with me. I would like to explore the counter-intuitive thought that nothing much happened when the universe came into existence. I am not denying that the Big Bang banged in its dramatic way: there is so much evidence in favour of it, and considerable evidence in favour of the inflationary era, that it would be absurd to reject it as an account of the primeval universe just under 14 billion years ago. I am suggesting reinterpretation.

The motivation for this view is a step towards confronting one of the great conundrums of existence: how something can come from nothing without intervention. One role of science is to simplify our understanding of Nature by stripping away misleading attributes. The awesomeness of everyday complexity is replaced by the awesomeness of the interconnectedness of an inner simplicity. The wonder at the delights of the world survives, but it is augmented by the joy of discovering the underlying simplicity and its potency. Thus, it is much easier to comprehend Nature in the light of Darwinian natural selection than simply to lie back and marvel at the richness and complexity of the biosphere: his simple idea provides a framework for understanding even though the complexity emerging from that framework may be profound. The wonder remains, and is perhaps intensified, that such a simple idea can explain so much. Einstein simplified our perception of gravitation by his generalization of his special theory of relativity: that generalization interpreted gravity as a consequence of spacetime being warped by the presence of massive bodies. His 'general theory' is a conceptual simplification even though his equations are extraordinarily difficult to solve. By weeding out the unnecessary and focusing on the core, science moves into a position from which it is more able to provide answers. To express the point more bluntly, by showing that not much happened at the Creation, it is more likely that science can resolve what actually did happen.

The weasel words in my stated aim, of course, are the 'not much'. Quite frankly, I would like to replace the 'not much' by 'absolutely nothing'. That is, I would like it to be the case that absolutely nothing happened at the Creation and that I can justify the claim. Gone activity, then gone agent. If absolutely nothing happened, then science would have nothing to explain, which would certainly simplify its task. It could even claim in retrospect that it had already been successful! Science has sometimes advanced by demonstrating that a question is meaningless, as in asking whether moving observers can agree on the simultaneity of events, leading to special relativity. Although it is not in the remit of science, the question of how many angels can dance on a pin head is eliminated if it can be shown in some manner or other that angels don't exist, or at least through some physiological or anatomical defect are incapable of dancing. So the elimination of a question can be a legitimate way to provide an answer. That might be a step too far, and perceived as a dereliction of scholarly duty, a cheat, a typical scientific cop-out—call the evasion what you will—for you to accept at this stage, so I shall confine my argument to an assertion that 'not much' happened when the universe came into existence, and in due course explain how much was not much.

* * *

The point of all this preamble is that the evidence I shall bring forward for not much happening is to show that the laws of nature stem from it. I shall argue that at least one class of natural law stems from not much happening on day dot. It seems to me that that amounts to powerful evidence for my view, for if the machinery of the world, the laws that govern behaviour, emerge from this view, then the need to have that elaborate hypothesis of a law-giving agent, commonly known as a god, is avoided. The laws that do not emerge from indolence I shall argue spring from anarchy, the imposition of no laws. In the course of planning this account, I came to realize that anarchy might in some cases be too restrictive, but then I saved the concept in all its potency by allowing anarchy to form an alliance with ignorance. You will come to see what I mean. At one stage I shall even invoke ignorance as a powerful tool for achieving knowledge.

I must stress that in this account I have in mind only physical laws, the laws that govern tangible entities, balls, planets, things in general, stuff, intangible radiation, fundamental particles, and so on. I leave aside moral law, which some still ascribe to a god promoted to God, the supposed inexhaustible, incomprehensible, free-flowing fount of all goodness, the arbiter of good and evil, the rewarder of the sheep and forgiver of the goat. My position, for the sake of clarity, is that biological and social phenomena all emerge from physical law, so if you were to mine down into my belief, you would find my view to be that an appeal to indolence and anarchy embraces all aspects of human behaviour. But I shall not develop that thought here.

* * *

So, what are laws of nature? What am I trying to explain by discovering their origin? Broadly speaking, a law of nature is a summary of experience about the behaviour of entities. They are sophistications of folklore, like 'what goes up must come down' and 'a watched pot never boils'. Folklore is almost invariably wrong in some degree. What goes up won't come down if you hurl it up so fast that it goes into orbit. Watched pots do boil eventually. Natural laws are typically improvements on folklore, as they have been assembled by making observations under controlled conditions, isolating the phenomenon they seek to explain from extraneous influence (the mud of Aristotle's cart, for instance, the air around his arrow).

Natural laws are believed to be spatially universal and timeless. The ubiquity and perhaps eternal persistence of laws means that any law of nature is thought to be valid not only from one corner of a laboratory to another, but across continents and beyond, throughout the universe. Maybe they fail in regions where the concepts of space and time fail, as inside black holes, but where space and time are benign, a law valid here and now is a law valid there and then.

Laws are established in laboratories occupying a few cubic metres of the universe but are believed to apply to the entire universe. They are formulated in a period comparable to a human lifetime but are believed to apply to something like eternity. There are grounds for these beliefs, but caution must be exercised in embracing them whole-heartedly.

On the little scale of human direct experience, a tiny fraction of the time for which the universe has existed and a vanishingly small fraction of its volume, laws have been found to be the same wherever and whenever they have been tested, on Earth at least. On the bigger scale of human experience they have been tested by the ability of astronomers to observe phenomena at huge distances from Earth, in other galaxies, and likewise far back in time. Unless distances in space and time are conspiring to trick us by somehow jointly cancelling deviations, no deviations from Earth-established laws have been detected. On the short timescale of a few billion years of past time there is no reason to suspect that for a similar period into the future our current laws will change. Of course it might be the case that over the next few trillion years, or even at midnight tomorrow, currently hidden dimensions lurking but suspected in spacetime will uncurl to augment our handful of familiar dimensions and transform our well-trodden laws into unrecognizability. That we do not know, but one day—such is the power of a law of nature—we might be able to predict on the basis of laws we are establishing now. Laws have within them the seeds of their own replacement.

Almost all, not all, laws are approximations, even when they refer to entities that have been insulated from external, adventitious influence (mud). Let me bring into view here one figure from history and the first of the little simple laws that I shall use to introduce a variety of points. (Later I distinguish between big laws and little laws; this is a little law.) The very clever, inventive, and industrious Robert Hooke (1635–1703) proposed a law relating to the stretching of springs.[2] As was not uncommon in that day, he expressed it as an anagram in order to claim priority but at the same time buying time to explore its consequences without fear of being outpaced by others. Thus, in 1660 he wrote cryptically and alphabetically *ceiiinosssttuv*, only later revealing that he really meant '*Ut tensio, sic vis*'. In the more direct language of today, his law states that the restoring force exerted by a spring is proportional to how far it is pulled out or compressed. This law is a very good description of the behaviour of springs, not only actual springs but also of the springlike bonds between atoms in molecules and has some fascinating consequences wholly unsuspected by Hooke or even by his contemporary Newton. However, it is only an approximation, for if you pull the spring out a long way the proportionality fails even if you stop before it snaps: *ceiiinnnoosssttuv*. Nevertheless, Hooke's law is a good guide to the behaviour of springs, provided it is kept in mind that it is valid only for small displacements.

There might be some laws that are exact. A candidate is 'the law of the conservation of energy', which asserts that energy cannot be created or destroyed: it can be converted from one form to another, but what we have today is in total what we shall have forever and have had since forever in the past. So powerful is this law that it can be used to make discoveries. In the 1920s it was observed that energy seemed not to be conserved in a certain nuclear decay, and one set of proposals revolved around the suggestion that perhaps in such novel and hitherto unstudied events energy is not conserved. The alternative view, proposed by the Austrian theoretical physicist Wolfgang Pauli (1900–58) in 1930, was that energy is conserved but some had been carried away by an as yet unknown particle. Thus was stimulated the search for and in due course successful detection of the elementary particle now known as the neutrino. As we shall see, the law of the conservation of energy is at the heart of the understandability of the universe in the sense that it lies at the root of causality, that one event can cause another, and therefore lies at the heart of all explanation. It will figure large in what is to come.

There are other laws that seem to have a similar status to Hooke's law (that is, are approximations and pleasant to know because they help us to make predictions and understand matter) and others that resemble the conservation of energy (are not approximations but lie deep-rooted in the structure of explanation and understanding). That suggests to me that there are two classes of laws, which I shall term inlaws and outlaws. Inlaws are the very deep structural laws of the universe, primary legislation, the foundation of understanding, the bedrock of comprehension. The conservation of energy is, in my view, an inlaw, and although I hesitate to say it, perhaps the mother of all inlaws. Outlaws are their minor relatives, like Hooke's and the others that will shortly come our way. They are secondary legislation, being little more than elaborations of inlaws. We can't do without them, and in many cases science has progressed through their discovery, application, and interpretation. But they are corporals in an army with generals at its head.

There is a special type of law that I need to acknowledge and draw to your attention: a law that applies to nothing at all yet is very useful. I need to unravel this perplexing remark. As I have already said, outlaws are typically approximations. However, in some cases the approximation becomes better and better as the material it is purporting to describe becomes less and less abundant. Then, taking this progression of diminishing abundance to an extreme, the law becomes accurate (maybe exact) when the amount of material it describes has been reduced to zero.

Here, then, is a so-called 'limiting law', one that achieves complete accuracy in the limit of having nothing to describe.

Of course, the way I have presented it makes it sounds as though the law is vacuous, applicable only to nothing. But such limiting laws are of enormous utility, as you will see, for in effect they scrape the mud off their own internal workings. Let me give an example to clarify what I have in mind.

The Anglo-Irish aristocrat Robert Boyle (1627–91), working in a shed just off the High in Oxford in about 1660 (where University College stands but possibly on land owned by my own college, Lincoln), acting perhaps under the suggestion of his industrious assistant Richard Towneley and in collaboration with the aforementioned intellectually ubiquitous Robert Hooke, investigated what was then regarded as the 'spring of the air', its resistance to compression. He established a law of nature that seemed to account for the behaviour of the gas we know as air.[3] That is, he found that for a given quantity of air, the product of the pressure it exerts and the volume it occupies is a constant. Increase the pressure, and down goes the volume, but the product of the pressure and the volume is a number that retains its initial value. Increase the pressure again, and down further goes the volume: and the product of the two retains its initial value. Thus Boyle's law (which the French call Mariotte's law), is that the product of pressure and volume is a constant for a given quantity of gas and, we would now add, at a set temperature.

The law is in fact an approximation. Squirt in some more gas, and the law is less well respected. Suck out some, and it gets better. Suck out some more, and it is better still. Suck out almost all and it is well-nigh perfect. You can see where this is going: suck out it all, and it is precise. Thus, Boyle's law is a limiting law, exactly applicable when there is so little gas present that it can be regarded as absent.

There are two points I need to make in this connection. First, we now understand (Boyle didn't: could not have because it depends on knowing about molecules, and that knowledge lay in his future) why the accuracy of the law improves as the abundance of the material declines. I won't go into it in detail, but essentially the deviations from the law stem from the interactions between the molecules. When they are so far apart (as in a sample that consists of only a wisp of gas, then these interactions are negligible and the molecules move chaotically independently of each other (the words 'chaos' and 'gas' stem from the same root: they are etymological cousins). The interactions are the internal mud that the reduction in the quantity of material wipes away to leave the clean perfection of chaos.

The second point is rather more important but closely related to the first. A limiting law identifies the essence of the substance, not the mud that its feet collect as it tramps through reality. Boyle's law identifies the essence of perfect gassiness, the elimination of the interactions between molecules that confuse the issue for actual gases in real life, so-called 'real gases'. A limiting law is the starting point for understanding the nature of the substance itself free from any accretion of details that distract and confuse. Limiting laws identify the perfection of behaviour of materials and will be the starting point for a number of our investigations.

Another important initial aspect of natural laws is that some are intrinsically mathematical and the others are adequately verbal. When I need to show an equation to substantiate a point, I shall send it to the Notes at the end of the book, to make it available for those who like to see the mechanism at work beneath the words. The advantage of a mathematically expressed law (Einstein's of general relativity, to choose an extreme example) is that it adds precision to the argument. In its place I shall make every effort to distil the essence of the argument. Indeed, it can be argued that the extraction of the verbal content, the physical interpretation, of an equation is an essential part of understanding what it means. In other words, a possible view is that not seeing the equation is a deeper form of comprehension.

Not all natural laws are mathematical, but even the ones that aren't acquire greater power once expressed mathematically. One of the deepest questions that can be asked about natural laws, apart from their origin, is why mathematics appears to be such a perfect language for the description of Nature. Why does the real world of phenomena map so well on to this extreme product of the human mind? I have explored this question elsewhere, but it is so central to our perception and comprehension of the world that I shall return to it later (in Chapter 9). I suspect that all truly deep questions about the nature of physical reality (the only reality apart from the inventions of poets), such as the viability of mathematics as a description of Nature, probably have answers that are bound together in a common source and need to be considered in the round.

* * *

Finally, I need to say a few words about a variety of handmaidens of the laws of nature. As I have said, a law of nature is a summary of observations about the behaviour of entities. There are then two steps in accounting for a law. First, a *hypothesis* may be proposed. A hypothesis (from the Greek word for 'foundation', in the sense of groundwork) is simply a guess about the underlying reason for the observed behaviour. That guess might receive support from other observations and so gradually mature into a *theory* (from the Greek for 'speculation allied with contemplation'; the word shares its origin with 'theatre'). A theory is a fully fledged hypothesis, with foundations perhaps embedded in other sources of knowledge and formulated in such a way as to be testable by comparison with further observations. In many cases the theory suggests predictions, which are then subjected to verification. In many cases, the theory is expressed mathematically and its consequences are teased out by logical deduction and manipulation (and interpretation) of the symbols. If at any stage a hypothesis or a theory conflicts with observation, then it is back to the drawing board, with a new hypothesis being required and then developed into a new theory.

Although I have set out this procedure—the cycle of observation followed by hypothesis maturing into theory, tested against experiment—as a kind of algorithm that scientists follow, the practice is rather different. The scientific method is a liberal polity and the gut plays an important role in the early stages of comprehension. Scientists have hunches, make intellectual leaps, certainly make mistakes, nick ideas from others, muddle through, and then just occasionally see the light. That is the actual scientific method, despite the idealizations of the philosophers of science. Their idealization is like a limiting law, identifying the essence of the scientific method stripped of its human mud, a human activity practised in the limit of the absence of humans and their frailties. The central criterion of acceptability, though, is maintained within the stew-pot of procedures: there is almost invariably the comparison of expected outcome with experimental observation. As Max Planck once said, 'the only means of knowledge we have is experiment: the rest is speculation'.

In some cases there are sharp deviations even from the ideal path. One of the most powerful theories in science is Darwin's theory of evolution by natural selection. There is no mathematics in his formulation of this extraordinary theory, but its power is augmented by the later mathematical elaboration that has occurred. I am also not sure that the theory is based on a law of nature. It is certainly based on observation, of the fossil record and the diversity of species, but the observations are perhaps too diverse themselves to be summarized by a succinctly stated law other than perhaps Herbert Spenser's aphoristic 'the survival of the fittest' (in his *Principles of biology*, 1864) or less contentiously and pungently that 'organisms proliferate that achieve reproductive success in the niches currently available to them'. As a law, it would certainly be a candidate for being classed as a mighty inlaw.

* * *

That is the background for what is to come. I have argued that there are inlaws and outlaws: hugely important laws and lesser laws, respectively. My first task is to identify and examine some of each variety and wonder whence they originate. I have asserted that I shall look for their origin in the indolence of nothing much happening at the Creation, and where that fails, then in the anarchy that followed. In due course I shall have to take you on a flight of speculation to discover the source of the relevance of mathematics to physical explanation of the real world, but that is distantly down the line.

2

Much Ado about Nothing

How laws might emerge from nothing

Nothing is extraordinarily fruitful. Within the infinite compass of nothing lies potentially everything, but it is an everything lurking wholly unrealized. Such remarks, of course, are deliberately enigmatic, for at this stage I want to catch your eye and pique your curiosity. They are akin, perhaps, to Hindu philosophy's unarguable but unsatisfactory definition of being as the absence of non-being. To avoid being tarred with the same brush, I need to develop the remark to illustrate the fructiferous potency of nothing, to see that contemplating nothing is not an exercise in vacuity, more akin to theology than physics, to show that testable conclusions can be drawn from nothing within the framework of science, and for you to believe that the enigma can be unwrapped and its content rendered meaningful. I want to show that nothing is the core idea, the foundation of the possibility of understanding the laws of nature and therefore everything that is and everything that does. In short, I want to show that nothing is the foundation of everything.

To lead you into understanding nothing at all and its consequences, and to help you to understand why indolence plays such an important role in establishing the mechanistic infrastructure of the world, initially I would like you to think of nothing in a very commonsensically primitive way. That way in due course will have to yield to something more sophisticated, and then common sense will have to wither a little, but for the start of this journey you can quite safely think of nothing as being empty space. Until I advise you otherwise, just lie back and think of miles and miles of uniform, empty space and of years and years of time stretching from the distant past and into the unfathomable future. Think of timeless uniform emptiness everywhere and everywhen.

Into this prairie-like image of flat spatial and temporal barrenness I need to introduce a single figure. The extraordinary and imaginative German mathematician Emmy Noether (1882–1935) was born in Erlangen, taught (bearing the brunt of the misogynistic attitudes of the time) in Göttingen, and then later escaped Nazi persecution to Bryn Mawr College in Pennsylvania. There, too young, she died, leaving a rich legacy of abstract mathematical concepts and theorems. Noether has been called (by Norbert Wiener, himself a noted mathematician, in 1935) 'the greatest woman mathematician who has ever lived' and she enthralled Einstein. She is absolutely central to my argument, and to theoretical physics, by virtue of a theorem she developed in 1915 and published a couple of years later. I cannot of course repeat her technical argument here, but her conclusion is very simple. She established that *wherever there is a symmetry, there is a corresponding conservation law.*[1] I shall unwrap this remark and explain what is meant by a conservation law, by symmetry, and the relation she established between them.

By a 'conservation law' I mean a law of nature that asserts that although events take place, a certain quantity remains unchanged ('is conserved'). I have already mentioned one such law, the conservation of energy, and initially I shall focus on it again here.

* * *

Energy is one of those concepts used widely and often in everyday discourse but which is very hard to pin down and say exactly what it is. Everyone buys a lot of it, but would be hard pressed to say what they had bought. The term entered physics early in the nineteenth century and was found to be so valuable that it swept through science, displacing arguments like Newton's who focused on the more tangible concept of force. Its introduction and the realization of its potency even caused whole textbooks to be rewritten. Force is almost literally tangible; energy is abstract. That is the source of its importance, for abstract concepts are in general more widely applicable than concrete concepts. Abstract concepts are the conceptual skeleton to which different observational flesh may be attached; concrete concepts are intellectual islands.

The etymology of the word 'energy', from the Greek words for 'work inside', is a clue to its meaning. Energy is the capacity to do work. That 'operational definition' might not give you deep insight into what energy actually is, but at least it lets you know how to recognize it: work is readily recognizable because it puts us back into contact with the tangible concept of force. Work is the process of achieving motion against an opposing force, as in raising a weight against the pull of gravity or using a battery to push an electric current through a circuit. The more energy that is available, the greater is the amount of work that can be done. A coiled spring possesses more energy than when it is uncoiled: when coiled it can do work, uncoiled it can't. A tank of water possesses more energy when it is hot than when it is cold. Some sort of engine could be contrived to do work by using the energy stored in the hot water, but not after the water has cooled.

There are various kinds of energy. They include kinetic energy, the energy due to motion, as in the motion of a ball; potential energy, the energy arising from location, as in the energy due to the gravitational pull of the Earth on a weight; and radiant energy, the energy transported by radiation, as in the warmth we feel brought to us from the Sun and its driving of photosynthesis and its cascade of consequences we call the biosphere.[2] Each type of energy can be converted into any of the others. Nevertheless, it appears to be a strict law of nature that *the total quantity of energy in the universe is constant*. If energy in one form is lost, then it must be present in perhaps another form or in the same form elsewhere. A familiar example of this constancy is when a ball is thrown into the air. Initially after being released it has a lot of kinetic energy. As it rises against the pull of gravity, its potential energy increases and its kinetic energy falls. At the top of its arc it is momentarily stationary. Its kinetic energy is then zero and all its initial energy is now present as potential energy. As it falls back to Earth, accelerating as it goes, its potential energy falls too and its kinetic energy increases. At each stage of its flight, from beginning to end, its total energy, the sum of its kinetic and potential energies, is constant. The law of the conservation of energy summarizes this constancy by stating that *energy can be neither created nor destroyed*.

In Chapter 1 I mentioned that so powerful is the law of the conservation of energy that it has been used to predict the existence of new fundamental particles by noticing where it has apparently failed. The Danish theoretical physicist, and originator of an early version of quantum mechanics, Neils Bohr (1885–1962), had wondered when considering some puzzling observations whether in certain newly investigated nuclear processes it had failed. But in fact it had not, for energy had been carried away by a previously unknown particle, the neutrino, the 'little neutral one'. The existence of the neutrino was proposed by Wolfgang Pauli in 1930, then hunted down and ultimately found in an experiment conducted in 1956.[3] This episode suggests that the law is not unlike David Hume's attitude to miracles, that it is more reasonable to disbelieve the reporter than to believe the reported. Thus, scientists treat any report of the failure of the conservation of energy with extreme scepticism. As in this instance, the law was questioned, probed, and passed. It is, of course, not inconceivable that there are dragons lurking in the unexplored parts of the cosmos and that in events yet unknown the conservation fails.

In due course, in Chapter 8 in fact, I must return to this point in connection with that great and widely misrepresented clarifier of human thought, Heisenberg's uncertainty principle, which some consider opens a loophole, allowing energy to fluctuate on very short timescales. On a broader scale, and more conventionally, the conservation of energy underlies the impossibility of perpetual motion, the performance of work without the consumption of fuel. Indeed, one component of the evidence for the law is the failure to achieve perpetual motion despite increasingly desperate attempts to do so. In fact, despite the repeated claims of charlatans, perpetual motion has never been achieved and is now regarded as unachievable. This observation, in a sense resignation from the hope of the prospect of infinite energy without effort, and as a consequence the abnegation of the prospect of infinite wealth, is one of the foundations of one great group of natural laws, the laws of thermodynamics, to which I return in Chapter 5. The successors of those charlatans, spurred on by the sharp end of the carrot of the prospect of riches galore, have of course persisted in presenting elaborate machines purporting to produce work from nothing, but all they have achieved is exposure, derision, and greater confidence in the law. To some extent we should be grateful to them (and certainly to their honestly toiling counterparts who took the trouble to overthrow the false claims), for the failure of persistent, aggressive, driven attacks on the law has led to the strengthening of the acceptance of its validity.

There is plenty of other evidence too, for all the calculations of the motion of particles based on Newtonian mechanics rely on it, and although there are deviations from Newton's predictions in certain cases, those deviations are due to other and well-known causes. The calculations of quantum mechanics also rely on its validity, and are invariably successful in the sense of giving very precise agreement with observation. There can be little room for doubt that energy is conserved, and conserved exactly.

The conservation of energy, although clearly of immense technological and economic importance, and for working out problems in textbooks, is actually even more important than it might appear. It is the foundation of 'causality', the seemingly undeniable observation that every event is caused by a preceding event. Without causality, events would be capricious and the universe would be a midden of disconnected happenings. With causality there is the prospect of comprehension in terms of tracing cause to its effect and effect to its cause. Causality brings the prospect of discovering order and systematic behaviour. That systematic behaviour is expressed by the laws of nature, and therefore allows the emergence of the form of comprehension embodied in science. The conservation of energy plays a central role in causality by imposing a powerful constraint on what can happen: energy must be conserved in any event. The conservation of energy is like a serious, watchful, and incorruptible police force, forbidding straying from the law that constrains energy to a single, fixed, and presumably cosmically unchangeable value. If energy were not conserved, there would be less constraint on actions that had been caused by a prior event and we might as well not have causality. Yes, there are other constraints too, but energy is so central to behaviour and universally applicable that its conservation is of paramount importance. As I remarked in Chapter 1, it is an emperor among laws, the mother of all inlaws.

* * *

So, why is energy conserved? What is the origin of this most imperial of laws? This is where Noether steps in and illuminates with her brilliant theorem the barren emptiness that I have invited you to consider. The core point, according to her proof that the conservation of something springs from the symmetry of a related something, is that the specific instance of *the conservation of energy*, our current focus, *springs from the uniformity of time*. This uniformity is the 'symmetry' aspect of Noether's theorem in this context.

What does this uniformity mean in practice? On the surface, uniformity in time simply means that you will get the same result if you perform the same experiment on Monday or Thursday (or whenever). That is, the swing of a pendulum or the flight of a ball is the same, provided all the conditions are the same, regardless of when you observe them. To express this uniformity, this independence of when they are applied, the laws of nature are said to be 'temporally invariant'. In practice, temporal invariance means that if you have an equation that describes a process at a particular instant, then the same equation applies at any other instant. In other words, the laws of nature don't change with time. The *consequences* of those laws might change, as a planet might have drifted into a slightly different orbit or you might have thrown the ball faster than you intended, but the laws themselves are invariant.

Now look under that surface. For the laws of nature to be temporally invariant, *time itself must progress uniformly*. That is, time can't slow down, then accelerate, then skid to a near halt a moment later. Think what that would mean for the flight of a ball, or on a bigger scale the orbiting of a planet, with time compressed for some of its path then extended for another: it is hard to imagine that a theory of the dynamics of its flight could have been constructed. The ball would seem to accelerate, decelerate, seem to hang in the air with no forces apparently acting. On Monday there would be one law, on Tuesday another. Even if time was not randomly capricious but undulated regularly, stretching and shrinking periodically, the flight of the ball would be puzzling and unlikely to be worked out even by a Newton and the world would be likely to remain a bewildering dynamical place. For the laws of nature to be independent of when they are applied, time must flow in a uniform way: tick, tick, tick . . . on and on, with a steady unbroken rhythm.

I can anticipate some of the arguments you might advance to undermine this justification of uniform time. One might be that our measuring instruments also stretch and shrink in synchrony with the time experienced by the ball. Then we might not, perhaps could not, notice its non-uniformity and somehow or other the physics of our measuring instruments (including our eyes and ears) would vary synchronously and we would be blind and deaf to the variation. I think that one rebuttal of this objection is that the equations we write down and solve to describe motion are not entities that stretch and shrink (in the sense of the 'time' that is present as a parameter in them does not vary in such a way) as we watch them, so in that sense they are an objective, not subjective, description of motion. Indeed, although the reverse of Noether's theorem (which would imply that if there is a conserved property then there must be a related symmetry) is not as secure as its 'forward' version (that if there is a symmetry, then there is a related conserved property), a further argument is that because we know that energy is conserved, then, with caution, we can infer that time must be uniform.

You might also object that when Einstein climbed on to Newton's shoulders, his vision of the cosmos revealed one in which time is distorted locally (that is the content of general relativity, the recognition of the distortion of spacetime by the presence of massive objects, such as planets), so time is not uniform locally, and therefore Noether's theorem is silent on the conservation of energy locally. That is an important objection and you are in good company when you raise it. It was, apparently, the suggestion by the extraordinarily insightful and influential German mathematician David Hilbert (1862–1943) that this objection needed to be considered that set Noether off in pursuit of her proof. She ended up with a supplementary theorem ('Noether's second theorem') to address the point. I shall evade the objection by advancing two excuses (which is always a suspicious procedure, in science as in life, for I have to admit that two excuses do not make an explanation).

Most importantly, and as in her original theorem, I shall confine my application of Noether's theorem to the global, to the entire universe. Even though when matter had formed and congealed into planets, solar systems, and galaxies, and distorted spacetime around them, overall, globally, there is uniformity, with stretches here compensated for shrinkages elsewhere. Viewed as a whole, spacetime, and its time component, is almost certainly flat. Secondly, any small enough region of spacetime is locally flat, and energy conservation applies to it.[4]

I hope you will now accept, with caution, that time is uniform on a global scale (and locally, in sufficiently small domains) and therefore that as a consequence (according to Noether's first theorem) energy is conserved. As I have remarked, if we could listen to the passage of time, then tick, tick, tick... would occur for ever. If instead time ran tick tick... tick... tick tick, and so on, time would not be uniform and as a consequence energy would not be conserved, the world would be incomprehensible, and science futile.

* * *

But why is time uniform? It is here for the first time in this chapter that I invoke indolence and my tentative suggestion that not much happened at the Creation. I need to take your mind back to the moment of inception of the universe, the moment of cosmogenesis, but before doing so there are several housekeeping points that I need to mention simply in order that they are set aside before you raise them.

First, it is conceivable that this universe might be the daughter of a preceding universe, which may itself be the daughter of a grandmother universe, and so on far into the backward abysm of time. However, presumably long ago there was a first universe—let's call it the Ur-universe—that emerged from absolutely nothing. This, our current universe, might be that Ur-universe and has itself sprung from nothing (and might go on to have, and perhaps already has had, progeny). My focus is on the Ur-universe whether it be this or an ancestor of this universe. The point is that at some stage it seems as though there must have been an event in which nothing turned itself into something, even if that event was some number of generations of universes ago. Even if that number of generations is infinite, it is still possible for the Ur-universe to have emerged a finite time ago.[5] I discount it here, having no evidence one way or the other and my gut is silent on the question, and irrelevant. Nor does it really matter for the current account.

Secondly, I might be being simplistic in extrapolating time back so far. It might be that time is one great circle of being, bending back on itself so that, like the surface of the Earth, there was no beginning. Far into the future we might discover ourselves far back in the past with the present, perhaps in a modified form, yet to come. We currently have experience of a few billion years, and have formed the view that time essentially migrates forward in a straight line. We have no evidence either for or against the possibility that that straight line is a tiny fragment of a vast circle. It is the current fashion to deride the Flat-Earthers who were blind to the curvature of the Earth. In due course we the Flat-Timers might be the object of similar derision. In short, I am perhaps being naïve in supposing that a beginning, in any sense, occurred. Here too my irrelevant gut is silent and all I can do is to point to the possibility but set it aside. The possibility, though, exists, although there is no evidence for it, and might prove to be an example of what I mentioned in Chapter 1, namely science's ability to achieve progress by demonstrating that a question is meaningless. In this instance, if time is circular, then there would have been no identifiable beginning, for where does a circle start? In that case science will be able to claim something of a Pyrrhic victory, for it will be able to claim that it has eliminated the puzzle about what went on when it all began, because it did not begin. Put another way: its beginning is in its end.

Circular time is not the only possible problem. At very short times after the incipience of the universe, either Ur or not, and on very close examination now, the concept of time might crumble. There are a number of ways in which this might take place. One possibility is that at extremely short distances, at the so-called 'Planck length', the presumed and familiar smoothness of space breaks down and the distinction between space and time is lost.[6] Contemporary physics is also lost, and no one has any idea, yet, about how to cope with this scenario. On that scale, spacetime is no longer a smooth, continuous fluid-like medium but more like a box of sand or a foam. This too is a problem I have to put on one side.

* * *

In the absence of any compelling view to the contrary, let's suppose that there was a beginning and focus on the moment when nothing became something, that focus of so much philosophical mythological, and theological speculation (and here, I have to admit, quasi-scientific speculation). Up to now I have encouraged you to think of nothing as merely empty space and empty time, in short, as empty spacetime. I now need to wean you away from that primitive view. From now on, by nothing I shall mean absolutely nothing. I shall mean less than empty space, I shall mean less than a vacuum. If you like, the Hindu absence of being. To emphasize the absolute emptiness of the nothing I want you to have in mind I shall call it Nothing. This Nothing has no space and no time. This Nothing really is absolutely nothing. A void devoid of space and time. Utter emptiness. Emptiness beyond emptiness. All that it has, is a name.[7]

At the inception of the universe (specifically, the Ur-universe, but I shall stop using that term for the sake of simplicity), this Nothing rolled over into something and our incipient universe became equipped with space and time. The immediate consequence of that rolling over is what we call the Big Bang, but I want, at this stage, to avoid the impression of banginess and would like to regard the Big Bang as an event taking place later than this rolling over. The rolling over in due course enabled the Bang in some sense. Science is currently silent on the mechanism that led to Nothing rolling over into something, and might forever be, although there have been speculations. Those of a more religious or even poetically secular inclination can perhaps be content with a vision of a creator, itself outside Nothing, putting its shoulder against Nothing, perhaps accidentally bumping into it, and causing it to roll (and perhaps, if accidental, now aghast at the consequences); but that is not science's way.

Let's focus on the rolling and set aside the question of how it occurred, leaving that important question to another day.[8] My speculation, as you know, is that not much happened when Nothing became something. You can probably accept that Nothing is absolutely uniform: it cannot have lumps, gaps, pits, stretches, and squashes, for otherwise it wouldn't be Nothing. Then, when the universe rolls into existence, with not much happening, it seems plausible to me that that uniformity of Nothing is preserved. Therefore, space and time, when they emerge from this event, are uniform. Time in particular is uniform, and therefore (says Noether) energy is conserved. Thus emerges this primary law of nature, followed by causality, science, and the comprehensibility of the physical world. It also opens up the prospect of the emergence, in due course, of agriculture, warfare, its microcosm sport, and those titillations and stimulations of the senses and the intellect, literature, music, and the visual arts. That 'not much' was wonderfully pregnant.

The view that 'not much' happened at the inception of the universe is admittedly a hypothesis, a speculation, that seems to go against the grain of reality. But hypotheses gain validity when it turns out that they have consequences in agreement with observation. The hypothesis that 'not much' happened has, in this instance, resulted in a verified observation, so perhaps it is not incorrect. Being not incorrect is, however, by no means a guarantee that it is correct, for alternatives might lead to the same conclusion. Science is harsh: a single consequence in conflict with observation is enough to condemn a hypothesis to history's swelling scrapyard of junk ideas that have accumulated over the ages; a single verified consequence is merely an encouragement to persevere with a quest, not a guarantee of validity. Hypotheses mature into theories as their implications, and in some cases predictions, multiply and conform to observation, but even a theory that has survived into contented middle age can be condemned by one false consequence.

All ideas in science live on the edge. One successful consequence is certainly, perhaps merely, satisfying, but are there other consequences of my speculation that are supported by observation and help to keep it at least temporarily secure from science's Death Row?

* * *

I have used space to discuss time; now I shall take time to discuss space. The space that is our home and sandpit of action came into existence at the incipience of the universe. In the same spirit as I have just argued, as Nothing rolled over into being something, that Nothing's uniformity was inherited by the emerging new born space. The symmetry invoked in Emmy Noether's insight into the connection between symmetry and conservation in this instance is the inherited uniformity of space.

What is meant by the uniformity of space? In the same spirit as the interpretation of the uniformity of time that I mentioned earlier, the interpretation of the uniformity of space means that an experiment done here will have the same outcome as an experiment done somewhere else. Experiments in different laboratories will have the same outcome. The laws of nature do not depend on where you are. The consequences of those laws might be different, for the conditions might not be identical, but the laws themselves are the same. Thus, although the law governing the swing of a pendulum is the same, the same pendulum swings with different periods at sea level and up a mountain where gravity is weaker. If you migrate from one location to another, you don't have to change the equation expressing the law. Laws are spatially uniform.

Because there is no intrinsic distinction between space and time (according to relativity, they are two faces of spacetime), any remarks that apply to time also apply to space. As for time, for a law not to be different from place to place, space itself must be uniform. That is, space is not squashed up here, spread out there, and so on. As for a similar distortion of time, it is hard to imagine that a theory of the dynamics of the flight of a ball could have been constructed if that were not the case. The same caveats apply as for time, with our attention directed at the overall uniformity of the universe or some little locally flat patch, so that we can use Noether's first theorem (the one connecting symmetry and conservation) rather than resorting to her second theorem (the one relating to distorted spacetime). Globally, space, I presume, is flat.

Noether's theorem now strides into the arena in the context of the uniformity of space. According to her theorem, an implication of the uniformity of space is the conservation of 'linear momentum'. I need to say a word or two about the concept of linear momentum and its conservation.

Linear momentum is the product of a body's mass and its velocity.[9] So, a heavy cannon ball moving fast has a high linear momentum. A light tennis ball moving slowly has a low linear momentum. Despite their synonymity in everyday speech, in science velocity is not the same as speed, for as well as depending on the rate at which position changes (speed) it includes the notion of direction. Thus, a body that is travelling at a constant speed but is changing its direction (for instance, a planet in orbit around the Sun) is continually changing its velocity. A ball hit by a bat might travel off back with the same speed as it arrived, but its velocity, and therefore its linear momentum has been reversed. When thinking about linear momentum, always think direction as well as rate. That slightly complicates the notion of the conservation of linear momentum, the law that the total linear momentum is unchanged in any event, because you have to keep in mind various directions of travel (the issue didn't arise when considering the conservation of energy, where direction plays no role), but it remains quite easy to visualize.

Linear momentum is conserved in the collisions of particles. A simple example is that of two identical billiard balls rolling towards each other at the same speed. The total momentum is zero (because the velocities of the two balls are equal but opposite, so sum to zero when added together), and when they meet both are brought to a stop, their total linear momentum then still being zero. If they are moving towards each other at an angle, their total linear momentum is no longer zero. When they collide they roll away from each other. It turns out that their new paths are such that their total linear momentum is unchanged. Whatever the angle of collision, whatever their possibly different masses, whatever their initial possibly different speeds, and however many balls are involved in a collision the total linear momentum after a collision is the same as it was before the collision. Linear momentum is indeed conserved, and is conserved on account of the uniformity of space.

The conservation of linear momentum that is predicted by the hypothesis of 'not much' happening in combination with Noether's theorem is a verified and far-reaching observation. It underlies the whole of Isaac Newton's system of mechanics, which is now referred to as 'classical mechanics'. Thus, it underlies the trajectories of bodies and the changes that take place when they collide or influence each other by the forces they exert, it underlies the exertion of pressure by a gas as its molecules collide with the walls of its container, it underlies the classical description of the planets, the stars, and the galaxies. It also underlies the operation of jet and rocket engines.

An alliance of Nothing and indolence has brought us a long way. It has provided the infrastructure of causality, by accounting for the conservation of energy, and it has begun to accommodate the emergence of activity in the arena of spacetime by providing an explanation of the underlying assumptions of Newtonian physics. Could there be more? Are there any other consequences of primordial indolence?

* * *

I have discussed *linear* momentum, the momentum associated with bodies travelling through space with their various masses and velocities. I need to point out that there is another kind of momentum, namely 'angular momentum', the momentum associated with rotational motion. The spinning Earth has angular momentum by virtue of its diurnal motion around its poles, its spin on its axis. It also has angular momentum by virtue of its annual motion in orbit around the Sun. All rotating bodies possess angular momentum even if they are not travelling through space. The Moon has its menstrual angular momentum as it orbits the Earth; it also has its own spin angular momentum as it rotates at a matching rate on its axis (and so presents the same face to the Earth throughout the lunar month). Now, another well-established law of nature is that, like linear momentum, *angular momentum is conserved*. Although you might cause a body to rotate, like spinning a wheel or contriving to spin a ball, and thus endow it with angular momentum, elsewhere a compensating angular momentum will have been produced. Every time you cycle off to the East, you decrease the angular momentum of the Earth on its axis by a matching but wholly negligible amount. Though negligible, the angular momentum police notice it. Every time you cycle off to the West you increase the spin of the Earth and the day shortens, wholly unnoticeably, except to the angular momentum police, who tolerate no transgressions.

Just as in the case of linear momentum, when assessing and understanding the significance of the conservation of angular momentum speed must be considered in conjunction with direction. But what is meant by the direction of rotational motion, for the direction of travel of the planet, or of any rotating object, is changing as it migrates along its path? To ascribe direction to rotation think of the motion as taking place in a plane, then attach an arrow at the centre of the circular motion and perpendicular to the plane. If the motion is clockwise as seen from below, then the arrow points up from the plane. If the motion is clockwise as seen from above, then the arrow points down from the plane. Think of the convention as being like an ordinary corkscrew: as the corkscrew is rotated it travels in the direction of the arrow in our convention—a corkscrew burrows into the cork as you turn it clockwise. When you travel forward in a car, all the wheels rotate clockwise when seen from the right-hand side of the car, so each wheel can be thought of as having an arrow attached to its hub and pointing leftwards. As the car accelerates, these imaginary arrows lengthen in proportion to the increasing speed. As you brake, they shrink. If you stop, then reverse, they emerge and grow rightwards. Images of Boadicea and her chariot come to mind: she would scythe her enemies to her left as she travelled forward, and scythe her enemies to the right as she travelled in reverse.

There is a further point that you need to have in mind. Linear momentum is related to velocity through multiplication by mass, a measure of the resistance to changes in linear, straight-line motion. The greater the mass, then the greater the inertia is to enforced change. The resistance to changes in angular momentum is similarly related to a quantity known as the 'moment of inertia', a name intended to convey not a brief catnap but the sense of resistance to rotational rather than linear motion. (The term 'moment' in physics is used to denote a kind of leveraging influence rather than an influence along a line of flight; like using a spanner to tighten a nut, which exerts a forcing moment, a torsion.) Two bodies could have the same mass but different moments of inertia. For instance, think of two wheels of the same mass but in one the weight is concentrated close to the axle and in the other it is concentrated close to the rim. It is easier to accelerate the former into rotation than the latter, and so it has a smaller moment of inertia. A body rotating rapidly with a big moment of inertia (like a flywheel) has a higher angular momentum than one with a small moment of inertia rotating at the same rate. Flywheels do their job of sustaining smooth rotational motion simply because, with their high moments of inertia, they are difficult to stop rotating. The magnitude of the linear momentum of a moving body is the product of its mass and the rate at which it is changing position along a straight line. By analogy, the magnitude of the angular momentum of a rotating object is the product of its moment of inertia and the rate at which it is rotating.[10]

The law of the conservation of angular momentum expresses the experimental observation that *angular momentum can be neither created nor destroyed*. It can be transferred from one object to another, as in the collision of two spinning balls or when you accelerate on a bicycle. But the total angular momentum of all the bodies in the universe is a constant (and probably zero). Angular momentum is conserved. If one object acquires angular momentum through the exercise of some accelerating torsion, then another linked object loses it. If in a collision, a ball goes spinning off, elsewhere there is a matching, compensating change in an object's (for instance, the Earth's) angular momentum. Perhaps the most visual image is that of a spinning ice skater, who spins faster as he (or she) reduces his moment of inertia by drawing in his arms, but maintains his constant angular momentum.

And why is angular momentum conserved? You now know that to understand the origin of any conservation law you have to use Noether's theorem and look for the underlying symmetry associated with the conserved quantity. In this case her theorem identifies the relevant symmetry as the 'isotropy' of space as the source of the law. By isotropy (from the Greek words for 'same turning') is meant the uniformity of space on travelling around a point. Think of a point, move out a little way, then travel in a circle around the original point. If you find that there is no variation in space (whatever that means) on the circuit, then it is isotropic in the vicinity of the point you had chosen. A table-tennis ball is isotropic, so is a golf ball if you ignore the dimples. Therefore, to account for the conservation of angular momentum you need to sniff out the reason why space has rotational uniformity.

By now you can probably see how the argument runs. You need to think about Nothing again. Nothing is isotropic. That is, the absolute nothing that preceded our universe (or the Ur-universe) must be isotropic. Were it not, and had lumps and dips, it wouldn't be Nothing. When Nothing rolled over into becoming something, not much happened (so runs my hypothesis). Thus, the isotropy of Nothing was preserved as space and time emerged and consequently the space we now possess is isotropic. The implication of that isotropy is that angular momentum is conserved. Yet another basic law of nature has emerged, without it being necessary for it to be imposed.

Incidentally, there is another aspect of Nothing worth mentioning. When we (here the we of whom I speak are collectively astronomers and cosmologists, our observers) look at the galaxies that throng the visible universe, not just the individual stars, we find that they rotate and therefore have angular momentum. However, the rates of rotation vary and the orientations of the arrows representing the directions of rotation are apparently random. When the total angular momentum of the visible universe is assessed (allowing opposite angular momenta to cancel), it turns out that the result is zero. Overall, the visible universe has zero angular momentum even though individual components rotate.

That is exactly what you should expect when Nothing rolled over into apparently something. Nothing (that is, Nothing; not merely a nothing masquerading as a Nothing at the start of a sentence) has no angular momentum, so it is hardly surprising that the emergent something inherits the same angular momentum. No angular momentum was created at the Creation; there is none (overall) today. Our current something has simply inherited the properties of its parent Nothing.

* * *

Where are we? I hope it is clear that the presumption that not much happened when the universe came into being has led to three major laws of nature: the conservation of energy and the conservation of two types of momentum—linear momentum and angular momentum. In each case, our universe seems simply to have inherited the uniformity of its precursor void and to display that inheritance as three major laws of nature. The laws didn't need to be imposed: they are simply the consequences of inheritance from a primordial precursor state, the state of absolute Nothing. There are other conservation laws (that of electric charge, for instance), and every one of them has an associated symmetry. To unearth them we need to look more deeply into the nature of Nature, and I shall return to the point in a later chapter. There are also a lot of little laws, what I have called the outlaws, and I have not yet explored their origin. Their time will come. For the moment it is time to find space for indolence to rest and for anarchy to rule.

3

Anarchy Rules

How some laws emerge from no laws

As in politics, so in science: when indolence lies exhausted from effort, anarchy enters and takes control. Once a universe has rolled into existence, preserving the various uniformities characteristic of Nothing and thereby giving rise to the great conservation laws of nature, some of the laws that then populate and govern it are the consequences of anarchy. In this chapter I argue that the classical mechanics developed in the late seventeenth century by Isaac Newton (1642–1727) and elaborated with such effect in the following two centuries, and the quantum mechanics of Erwin Schrödinger (1887–1961) and Heisenberg, are manifestations of anarchy. Schrödinger and Heisenberg's system of laws developed early in the twentieth century by them and elaborated by others underlie Newton's earlier extraordinary formulation. They remain deeply puzzling but are the result of Nature let loose to run wild except for the constraint of the great laws of conservation arising from primordial indolence. I have shown that indolence polices the arena; here I aim to show that from anarchy spring the laws of behaviour in that arena.

I intend to talk about both classical and quantum mechanics. I shall do so by describing their content in a very general way, presenting the handful of central concepts of both without (in the case of quantum mechanics) getting bogged down in the fascinating but vexed question of interpretation. In fact, interpretation is secondary to the laws that govern the behaviour of entities, for interpretation is but an attempt to establish a way of thinking about the consequences of the laws in terms of concepts related to everyday experience. The laws of nature, as distinct from the laws that govern societies, can exist without interpretation; but of course as I have already said, it is inherently a component of science to seek to interpret what the laws imply and thus to enlighten us about physical reality.

The interpretation of classical mechanics is easy because that description of behaviour is in direct contact with everyday experience through our perception and understanding of location and velocity. The revolution of thought associated with quantum mechanics has engendered a perplexity that remains unresolved even though the numerical and observational predictions—arising from the implementation of its laws (that is, just getting on with the calculations they entail), which is distinct from their interpretation—are invariably (so far, at least) in agreement with observation to a remarkable degree. It might be that human brains are ill-equipped to break from their inheritance of observing the motion of familiar objects and simply cannot come to terms with the counter-intuitive unfamiliarities of quantum mechanics. We have brains built—evolved for survival's sake—to cope with the savannah and the jungle, and there might be fundamental neurological structural reasons why we cannot understand quantum mechanics. One day in the future, I suppose, a quantum computer might claim to understand its own principles, leaving us still perplexed despite us having built it. The difficulty, perhaps intrinsic impossibility, of understanding quantum mechanics is for my purpose irrelevant. I am concerned with the laws themselves, not a human being's difficulty in grappling with what they mean.

My aim, then, is simply to show that a central plank of quantum mechanics emerges very naturally from anarchy and that that plank is a springboard for the emergence of Newton's classical mechanics in a very straightforward way. To do so, I need to take you first on what might seem to be a rural detour but which will turn out to be a highway to understanding. The detour will introduce you to an outlaw that will turn out to be the mother of an inlaw.

* * *

One little law of nature, the 'outlaw' in my classification, is that *light travels in straight lines.* Associated with that behaviour is the consequential outlaw that governs reflection in mirrors, where the angle of reflection is equal to the angle of incidence. That is, a flat looking-glass mirrors the world accurately: it doesn't distort the image and thereby render it unintelligible or at least deceitful. You might also remember 'Snell's law', discovered and rediscovered by a variety of people including, in 1621, the Dutch astronomer Willebrord Snellius (1580–1626), concerning the angle of refraction when a ray of light bends at the interface of air and water (or any other transparent medium) with the angle depending on the relative values of the refractive indices of the air and the medium (more on this shortly).[1] What is the origin of this behaviour?

In each case, and at a slightly deeper level of integration and understanding that combines both laws of reflection and refraction, the general law is that *light adopts the path between its source and its destination that takes the least time.* (Don't ask why yet: that will come later.) This is the 'principle of least time' proposed and explored by the French mathematician Pierre de Fermat (1601?–65), although it can be traced back to the Greek mathematician, inventor, and early practitioner of the scientific method of experiment in alliance with theory, namely Hero of Alexandria (*c.*10–*c.*70 CE) in about 60 CE. It can also be traced to the Arab mathematician, astronomer, and similar pioneer of the scientific method, Ali-Hassan ibn al-Hasan ibn al-Haythan (more conveniently, avoiding parentage and then Latinized, Alhazen; *c.*965–*c.*1040) in 1021 in his book on optics.

In passing, I would like you to bear in mind, for it will become relevant shortly, a similar proposal, the 'principle of least action', where 'action' has a technical meaning that I shall clarify later but which you can take at face value as 'effort' for now. This principle was proposed by the French philosopher and general all-rounder Pierre Maupertuis (1698–1759) in the early 1740s and is relevant to the propagation of particles (which include anything from pea to planet). According to it, *particles adopt paths that minimize the action associated with the path between two fixed points.* A ball in flight is tracing a path that once it leaves the bat minimizes the action. The Earth has sniffed out an orbit, given its orbital speed and distance from the Sun, that involves least action. The other planets and bits of cosmic debris have too, in their characteristic ways. A projectile in flight travels along a trajectory that involves the least action between its origin and its target. Any other path would correspond to a greater action, more effort. It is as though the principle of indolence is at work again, with least action corresponding to the greatest laziness; but I shall show that another principle is at work than this anthropomorphic allusion might suggest.

There is also a deep hint lurking in these two principles, one principle applying to light and the other principle to particles: perhaps because both principles have a similar form they have a similar origin and can be unified; I shall build on this hint in due course. A meta-principle of science is perhaps that coincidences in Nature are always worth exploring, for they might spring from structural analogies and point to deep relationships. Coincidences, which are always suspicious, when pursued can furnish insight. The early proponents of the principles were not entirely clear whether least time and least action were moral obligations imposed by an austerely commanding Nature. They are of course physical principles, and both arise, as I shall argue, from that least morally constrained of polities, namely anarchy.

* * *

Let's focus on the mirror again. It's a fairly straightforward matter of geometry to show that the shortest distance from source to eye via a flat mirror, given that both legs of the journey are straight lines, is one in which the angle of incidence is equal to the angle of reflection. That route also takes the least time for light travelling, as it does, at a constant speed. Therefore, according to the principle of least time, the path taken by light when it reflects from a mirror is such that the angle of reflection is equal to the angle of incidence.

Now to Snell and his law of refraction. Light travels more slowly through glass or water (or any other dense transparent medium) than it does through a vacuum or air. The ratio of its speed in a vacuum to its speed in the medium is called the 'refractive index' of the medium. For water, the refractive index is about 1.3, so light travels about 30 per cent more slowly in water than it does in a vacuum (or in air, which is much the same thing for our purposes here). You can probably wade through deepish water about ten times less speedily than you can walk in air, so if you ascribed to water a refractive index for your motion it would be close to about 10.

Now consider what the path will be that corresponds to the least time of flight when light leaves a source in air and needs to reach a destination under water. A similar problem is how you can get most quickly to someone drowning in a lake. Let's keep it to walking and wading just to keep the picture simple. You could take a straight line from your deckchair to the victim, which will entail a certain amount of walking and then wading. Alternatively, you could cut down the wading time to a minimum by walking across to a point immediately opposite the victim and then wade the short way in from the water's edge. The problem is that you have lengthened the time spent in walking because the path on land is longer than before and the brief wading episode does not fully compensate for that additional time of walking. As you probably suspect, there is an intermediate path that minimizes the total time: walk at an angle across the land then wade at an angle through the lake to the victim. The precise angle that you will have to turn as you enter the water and start to wade depends on your relative speeds in the two media and therefore on their refractive indices. In fact, as a little geometrical reasoning can show, the angles that you need to make as you run across the land and then wade through the water, and which light by analogy makes as it passes from one medium to another, is given by Snell's law of refraction.[2] From Snell's law then stems all the properties of prisms and lenses, which rely on refraction for their operation, and thence the whole of what is known as 'geometrical' optics. Geometrical optics is so called because it treats the paths taken by light as a succession of straight lines that make the appropriate angles to each other: the overall path is constructed geometrically.

At this stage in the argument, I have told you about two little laws, of reflection and refraction, and have shown that they are both manifestations of a deeper law, that light adopts the path of least time between source and destination. But why does light behave like this? What is the origin of this fundamental law of geometrical optics? And how does light know, seemingly in advance, what is the path corresponding to the time of least travel? Suppose it started off in what turns out to be the wrong direction, should it turn back and start again (which would add even more time) or just plough on and hope for the best? What is the origin of its apparent prescience?

* * *

The origin of the law of least time is anarchy. To see how that works, we need to build on the fact that light is 'electromagnetic radiation', a wave of oscillating electric and magnetic fields travelling at what unsurprisingly is called the 'speed of light'. To avoid this apparent tautology, I think that that speed ought to be called something like the 'Maxwell speed' after James Clerk Maxwell (1831–79), a life all too brief, but one of the immortals of science, the nineteenth-century Scottish physicist who first identified the electromagnetic nature of light.

To picture light, think of a wave as a train of peaks and troughs hurtling through space, so that each peak is travelling at the speed of light. The distance between neighbouring peaks is the 'wavelength' of the light. The heights of the peaks and depths of the troughs indicate the strength of the electric field at each point, and therefore the brightness of the light: a peak indicates that the field is directed up from the direction of travel, and a trough indicates that the field has changed direction and is pointing down. If you could hold up a finger and sense the passage of the ray, you would feel a series of pulses of electric field changing direction rapidly as the light passed you. The frequency with which the pulses change direction corresponds to the colour of the light: if the pulses change relatively slowly (but still actually very rapidly) the light would be red: it the pulses change more rapidly, then the light would be yellow, blue, or even violet. White light is a mixture of rays of all colours from red to violet. Infrared (below red) and ultraviolet (beyond violet) speak for themselves. Slow the alternation of pulses down, and you get radio waves; speed the alternation up, and you have X-rays. A final point in this connection is that, because all colours of light travel at exactly the same speed, the pulses change direction more rapidly (the light has a greater frequency) if the separation of the peaks (the wavelength) is short. Thus, blue light has a shorter wavelength than red light, and X-rays have a shorter wavelength still. The wavelength of visible light is about one half of one-thousandth of a millimetre, which—despite the claims of the pessimists who regard that as 'unimaginably small'— is perhaps almost within reach of imagining.

With that description and anarchy in mind, consider a wave of a single colour starting out from some source and taking an entirely arbitrary path to a destination (your eye, for instance), perhaps curling around, backing up on itself, then travelling directly into your eye. The wave is a sequence of peaks and troughs along this wayward path, and where it enters your eye let's suppose it happens to be a peak. Now consider a path that closely matches that first path but is not quite the same, a slightly different curl, a later dash to your eye. Once again the wave is the same series of peaks and troughs, but at your eye it might end up as being not a peak but a trough or something close to a trough. This trough cancels the peak from the first path (the electric fields they represent are in opposite directions and cancel each other). That cancellation might not be perfect, but you can imagine a large number of paths neighbouring the first path, each one having a slightly different length. Overall, the peaks and troughs of this multitude of neighbours will wash each other out at your eye. In other words, even though the paths have been allowed, you won't see light that has travelled by them. Anarchy has done away with itself.

Now consider a straight path between the source and your eye. Once again, the wave that travels along it might end up as a peak at your eye. Now consider the light travelling along a neighbouring path that is almost but not quite the same as the straight path. It will end up as almost a peak because the distance travelled is almost the same as along the straight path. In particular, because the paths are so similar, it doesn't end up as a trough to cancel the peak of the first path. There are many other paths that are very close to being straight, and they all end up as almost a peak. They coexist: they don't wash each other out. In other words, you will see light travelling by these almost straight paths. Anarchy has allowed a loophole.

It might seem that I have rigged the argument, and have asked you to take on trust that wiggly paths have destructive neighbours but straight paths don't. That is true, but I am basing this distinction on hard mathematical truths. I know that that sounds a little like saying 'trust me, I'm a doctor', but it is a standard result of conventional 'physical optics' (or 'wave optics', the version of optics that recognizes the wave nature of light rather than immediately treating the paths it takes as geometrically straight). The Notes sketch the bones of the relevant argument.[3]

You should now be able to appreciate where the discussion has led. No law has been imposed governing the propagation of light (light can take any path whatsoever between source and destination), but the outcome of that laissez-faire behaviour is a law (that 'light travels in paths such that its time of flight is a minimum'). Light doesn't know in advance what path will prove to result in the shortest time of flight; it has no prescience. It tries every path simultaneously, and it turns out that only the path of least time survives destruction by its neighbours. From anarchy has emerged law.

So far I have talked only about light travelling through a uniform medium, such as a vacuum or (to a good approximation) air. How does Snell's law emerge, and indeed the whole of the optics of the lenses in cameras and microscopes, where light is bent in a variety of ways when it passes through a succession of differently shaped transparent media? In these cases it is necessary to take the refractive index of the medium into account and its effect on the train of peaks and troughs. Light travels more slowly in a dense medium, so although the direction of its electric field alternates at the same rate (light doesn't change colour when it enters water or glass), the peaks and troughs of the wave are closer together: the wavelength is shorter. That modification of the spacing of the peaks and troughs turns out to modify the identity of the path that has non-destructive neighbours. The surviving path is no longer the straight-line path between source and destination, but a path that is a straight line in one medium then bends into another straight line in the second medium. When the analysis is followed through, it turns out that the surviving path, the one with neighbours that don't wash it out, is exactly the one that corresponds to least time of passage and given by Snell's law. Anarchy is the basis of all optics.

I need to interpolate an important remark. The washing out of neighbouring paths is more effective the shorter the wavelength (the higher the frequency) of the radiation. When the wavelength is very short, even tiny deviations of paths from each other gives very different relative locations of peaks and troughs and the cancellation is complete. The argument I have presented is less stringent when the wavelength is long, and then quite big deviations of the paths from one another need not result in their cancellation. The 'geometrical optics' I have referred to treats light as having such a short wavelength that even infinitesimal differences between paths are effective, with the result that light can be treated as travelling in a sequence of geometrically straight lines. In practice, light has a measurable wavelength and geometrical optics is an imperfect approximation. The resulting allowed small deviations from straight lines are observed as the aberration of lenses and various other optical phenomena. Radio waves have wavelengths of the order of metres and more, so for them geometrical optics is a poor approximation and even large objects do not block their path. Radios work round corners.

Although sound is not electromagnetic (it is a wave-like propagation of pressure differences), it follows the same rules of propagation as light. However, typical wavelengths are of the order of metres (the wavelength of middle C is 1.3 metres), so 'geometrical acoustics' would be a very poor approximation in a world packed with human-sized objects. That is why we can hear round corners.

* * *

You might, quite reasonably, object that the 'law' I have selected to demonstrate the self-constraining role of anarchy is rather trivial, a juvenile outlaw. As this chapter evolves, however, you will come to see that on the contrary it is awesomely rich and that there is more to this discussion of the propagation of light than meets the eye. Exactly the same anarchy that imprisons light by its absence of constraint imprisons matter likewise and thereby accounts for quantum mechanics and, by extension, even classical mechanics too.

The crucial concept I require in order to make the connection is that *particles are wave-like*. Here I need to lead you into the midst of another of those great revolutions of science that displaced common sense by something that in a number of respects turns out to be simpler even though it might seem shocking and unlikely at first sight. As I remarked in Chapter 1, classical physics kept particles neatly in one pile of thought and waves in another. Their characters were entirely distinct. Particles were point-like entities localized in space; waves were undulating entities that spread perhaps endlessly through space. What could be more distinct? Who could fail to distinguish one from the other?

Nature could. First, it dawned on scientists that certain experiments on the lightest particle then known, the electron, showed that it behaves as though it is a wave. I have already mentioned my image of J. J. Thomson and his son George at breakfast with an icy stillness between them. J. J. had shown incontrovertibly that an electron, which he had discovered in 1897, is a particle with a well-defined mass and charge. His son, G. P., had come along, and in 1927 had shown equally convincingly that an electron is a wave. At about the same time, the same conclusion had been reached by Clinton Davisson (1881–1958) and Lester Germer (1896–1951) who had shown that electrons undergo exactly the same type of 'diffraction' characteristic of wave-like light.[4]

There were other unsettling puzzles too. Everyone knew that light is a wave (I have used that concept in this chapter, of course), and myriad experiments over decades had confirmed it. Then two nuisances came along. The first nuisance was the German-Swiss-Austrian-American patent clerk Albert Einstein, who showed in 1905 that a certain effect involving the ejection of electrons from metals when illuminated by ultraviolet radiation (I am referring to the 'photoelectric effect') could be explained readily if light rays were actually streams of particles. (That explanation won Einstein his Nobel Prize in 1921; relativity, though intellectually richer and more far reaching, was held in greater suspicion at the time.) The particles in due course became known as 'photons'. The other nuisance was the American physicist Arthur Compton (1892–1962), who in 1922 showed that in order to account for the way that light bounced off electrons, the light had to be treated as a stream of photons acting like little bullets. (He got a Nobel Prize, too, in 1927.) Physics was in a quandary: what were known to be particles behaved like waves, and what were known to be waves behaved like particles. What on Earth was going on?

From the recognition of this dual character of the electron and of the photon, and later the dual character of all matter and radiation, emerged the concept of 'duality' in which entities show both particle-like and wave-like character depending on the type of observation being made. The recognition of duality is one foundation of quantum mechanics, which emerged under the hands (more specifically the brains) principally of Werner Heisenberg, isolated on an island; Erwin Schrödinger, on a mountain with a mistress; and Paul Dirac (1902–84), in world of his own, in 1925–7. Quantum mechanics and its fancy developments, such as 'quantum electrodynamics', have provided numerical predictions of extraordinary precision and have been tested seemingly exhaustively to umpteen decimal places, with no deviation or discrepancy ever detected. In other words, quantum mechanics, which is based on the recognition of duality, might be right.

With these preliminaries established, and the duality of matter and radiation recognized, it is now time to see how that concept opens the door to understanding the origin of the laws governing the motion of particles. Providing particles behave like waves, you already know the answer to the existence of rules: there are no rules. The waves travel between source and destination without constraint, taking all possible paths. However, all but one path, the path that takes least time (I need to gloss that, and will do so shortly), have destructive neighbours and are washed out. Only that special path has neighbours that do not destroy it, and that is the path we conclude the particle takes.

So, let's regard an intrinsic character of a particle (not just an electron) as being wave-like. It follows from anarchy that a particle will travel in a straight line through a uniform medium, just like light and in the light of the arguments I have described already. That is, allow a particle to travel by any route whatsoever through empty space, and by virtue of its wave-like character the only path that will have non-destructive neighbours will be the straight path between source and destination.

* * *

Motion in straight lines is not the only feature of particles, and the world would be a tedious, over-predictable place if that were all. What about more complex behaviour such as orbits and paths that bend: does anarchy control it too? Yes; by taking note of action.

I am in some difficulty here, for 'action' is a technical term that faintly reflects the everyday notion of the same name: I have already urged you to think of action as effort, exertion, physical puff. Of course in physics there is a more formal definition buried away in the equations of classical mechanics, but it would be too much of a technical marathon to introduce it here in its full glory.[5] So, I shall stick with action as denoting puff.

When a classical particle travels along its trajectory it involves a certain amount of puff. According to Maupertuis' principle of least action, the path that a particle actually adopts is the one corresponding to least puff, as we would, perhaps, ourselves seek to adopt. It follows that the paths of particles that would be calculated from Newton's classical mechanics according to his original formulation in terms of the forces acting at each point, can be calculated instead by working out what trajectory involves least action.

The question immediately arises as to how a particle knows in advance the path of least action. It might start off along a seductively inviting path but then run into a hill of opposing force and have to puff seriously to mount it. By then, it would be too late to turn back and start again along an easier, less puff-demanding route.

Exactly the same problem arose in the discussion of the propagation of light and exactly the same solution is at hand. In that case, the resolution was the wave nature of light in collaboration with anarchy. In this case, the resolution is the wave nature of matter in collaboration with anarchy. In place of the refractive index of the medium through which the light travelled and slowed its motion and therefore modified its phase (whether it is a peak, a trough, or something in between) at its destination, is the action along the trajectory (the puff associated with the path), which affects the passage of the wave through regions of different potential energy and thereby controls the phase of the matter wave at the destination. These features were explored by that great, imaginative, scientific explorer and expositor Richard Feynman (1918–88) who found that he could use them to arrive at quantum mechanics.[6]

Here is the argument, with the particle of interest treated as a wave and anarchy ruling by abandoning control. The waves travel along all possible paths between source and destination, each one ending up with a certain phase that depends on the action involved in travelling along the path.[7] In general, all the paths have neighbours that end up with a significantly different phase—some peaks, some troughs—and bring death to the original path. Only the path that corresponds to least action has benign neighbours that do not wash it away. We the onlookers see none of this underworld of conflict: we notice that the only path that survives is the path of least action.

If the action is very large, as it is for heavy particles like balls and planets, the cancellation caused by neighbours is stringently effective and, as in geometrical optics, the particles can be regarded as travelling along a precisely defined surviving trajectory. Classical mechanics thus emerges from the wave nature of particles just as geometrical optics emerges from physical optics and the wave nature of light. When the particles are small, like electrons, the action is miniscule and the cancellation by neighbours is largely ineffective, just like the propagation of sound, and classical mechanics fails, just as 'geometrical acoustics' is inapplicable to sound. Quantum mechanics, always present but concealed when the action is large, now has to be used to evaluate the motion of the particles.

* * *

Finally, to round off this discussion, I need to mention the concept of a 'differential equation', for many laws of nature, and especially those of classical and quantum mechanics, are expressed in their form.[8] The topic is important because it is often remarked that the core mathematical feature of physics is a differential equation. An ordinary, familiar equation tells you how one property depends on another, as in $E = mc^2$, which tells you how the energy E depends on the mass m. A differential equation tells you how the *change* in a property (hence the allusion to difference in 'differential') depends on various properties, including the property itself. Newton's second law, that *the rate of change of momentum is proportional to the force acting*, is an example of a verbalization of a differential equation.

An important additional point here, though, is that differential equations are expressions for *infinitesimal* changes in a property. The reason for (and advantage of) that restriction is that the conditions might change from point to point. For instance, the force in Newton's second law might change from place to place and from instant to instant, and to find out its overall effect on the trajectory of a particle it is necessary to consider the cumulative effect of a lot of little, in fact an infinite number of infinitesimal, steps. We say that the overall effect of a force must be found by 'integration' (effectively, combining all the little steps), or equivalently that the differential equation must be 'solved by integration'. Thus, if at one point in space a force has a particular influence and at a neighbouring point it has a different influence, there is a nudge at the first point and another nudge at the second point, and overall the outcome of the forces is the sum of the two nudges. As I hope you can now appreciate, a differential equation is used to evaluate how a particle (for instance) gropes its way towards a destination and thus traces out a trajectory.

The all-important, pervasive differential equations of physics are the children of anarchy. As you have seen, anarchy leads to the laws of least time in optics and least action in mechanics, but these minima refer to the whole path, not an infinitesimal fragment of a path. Nevertheless, a remarkable mathematical result is the fact that *groping forward in accord with the appropriate differential equation guarantees that you will find yourself moving along the globally minimal path.* All that a differential equation is doing is giving you instructions about what to do at each point and time, whether to shift to the left or right, whether to accelerate or not, and so on, to ensure that you end up at your destination having travelled along the path of least time or of least action.[9] A global criterion has been decomposed into a series of local instructions. Therefore, although differential equations are widely taken to be the core feature of physics, it is arguable (and I like to think true) that, in classical and quantum mechanics at least, they are secondary constructs and that the core features are in fact the anarchy-based global features of behaviour, with the differential equations merely showing how to behave in practice locally, a kind of hitch-hiker's guide to the trajectory.

* * *

Where has anarchy brought us? We have let loose light to find its own destiny without imposing a rule, and found that it settles down into obeying a rule, that it travels by the path of least time. We have accepted the experimental evidence for the duality of particles, in particular their wave-like character, and by letting them loose without imposing a rule, have found that they settle down into obeying a rule, that they travel by the path of least action. Just as geometrical optics emerges from physical optics as light's wavelength lessens, so classical mechanics emerges from wave mechanics (the older name for quantum mechanics), as the body in motion ceases to be tiny and becomes of a familiar everyday size. We have also seen that the centrally important, fundamental, and ubiquitous differential equations of classical and quantum mechanics are local instructions for finding and groping along the path that satisfies the global criteria of least time for light and least action for particles. Anarchy has brought us to physics.

4

The Heat of the Moment

Laws relating to temperature

If I were marooned on a desert island with only a palm tree for company and water, water everywhere, there is one concept that I would like to accompany me. It is extraordinarily rich in consequences, gives deep insight into the nature of matter and the transformations it undergoes, and elucidates one of the most elusive concepts of science yet familiar aspects of everyday life, namely temperature. I shall introduce you to that companionable concept shortly.

That temperature plays a role in the properties of matter and the laws that governs it earns it a place in this account. This chapter also serves an initial introduction to 'thermodynamics', the richly important body of laws relating to the transformations of energy, such as the relation between heat and work and why anything happens at all.

Temperature enters the description of the properties of matter in two ways, one from the world of observable phenomena, which we call the 'phenomenological aspect', and one from the underworld of atoms and molecules, which we call the 'microscopic aspect' or, because this world lies beyond the reach of conventional microscopes, the 'molecular aspect'. We are all familiar in a general way with temperature and the various scales on which it is reported and know its importance in our physiological well-being. We know that there are hot objects and cold objects, and we know that raising the temperature is essential for bringing about the changes that industry and chefs require. But what is it? And could temperature be an aspect of anarchy in alliance with indolence?

To deal with all these matters, I would like you to meet one of my heroes, a scientist well-known among scientists, but whose name does not spring lightly or commonly from the general public's lips. He is the Viennese theoretical physicist Ludwig Boltzmann (1844–1906), who, though short-sighted, saw further into matter than most of his contemporaries and, distraught at the unfavourable reception of his ideas, hanged himself. Boltzmann's ideas forged the link between the microscopic and the phenomenological, clarified the concept of temperature, and established a way of thinking about the properties of bulk matter in terms of the behaviour of its component atoms. His ideas explain why matter in the everyday world persists, and why matter when heated is liberated into the world of chemical change. A good *conceptual* companion, therefore, for those marooned on an intellectual desert island, but not perhaps a jolly one to be an *actual* companion on an actual isolated isle.

* * *

I have no idea whether Boltzmann thought in the way I am about to describe, and am in fact confident that it is unlikely he did, but here is a picture that captures the essence of his approach.

Imagine that you are lying in front of a many-shelved bookcase surrounded by piles of your books (for this discussion, they are all identical). Lying there, perhaps blindfolded, you idly toss your books on to the shelves. You take off the blindfold, and note the distribution of books over the shelves. There will be some on the high shelves, some on the middle shelves, and some on the lower shelves. There is no particular pattern. You clear the shelves, resume your position, and toss them on the shelves again. You open your eyes and see another seemingly random pattern. It is very unlikely that all the books will be on the high shelf or all on any one particular shelf.

You, and this vision of course is fantasy, repeat the procedure millions of times, noting the distribution after each episode. Some distributions of books (all on one shelf, for instance) occur hardly ever, others occur quite often. However, and this you note with interest, there is an overwhelmingly most probable distribution, one that turns up time and time again. In this distribution, most books turn up on the lowest shelf, fewer on the next higher, fewer still on the next higher, and so on, up to the highest shelf which might lack any books at all. This most probable distribution of populations of the shelves is the 'Boltzmann distribution', the core concept of this chapter and my favoured conceptual companion on the intellectual desert island.

Boltzmann's actual distribution applies not to books on shelves but to molecules and atoms. As is now well known, a consequence of quantum mechanics is that the energy any object can possess is limited to discrete values. A molecule cannot vibrate or rotate with arbitrary energy: it can accept energy only in steps ('quanta'). Even you on your bicycle accelerate in jerks, but the jerks are so tiny that for all practical purposes you accelerate smoothly. The jerks, however, are far from negligible for atoms and molecules. These 'energy levels', the allowed energies, are the bookshelves of the analogy. The books of the analogy are the atoms and molecules. Your nonchalant tossing is the random jostling that drives atoms and molecules from energy level to energy level. The outcomes of these random tossings at the shelves are the populations of molecules over their available energy levels. You are hardly ever likely to find all the molecules in the same energy level. The most probable distribution of molecules as a result of this random scattering of molecules over their available energy levels is the 'Boltzmann distribution', with most molecules in their state of lowest energy, fewer in the next higher level, still fewer in the next higher, and very few, perhaps none, in very high energy levels.

Now, I have to admit that the Boltzmann distribution isn't solely the outcome of random, anarchic behaviour. Indolence is involved too. The total energy of the molecules is fixed (this is a consequence of indolence and its implication, the conservation of energy, as I argued in Chapter 2). Therefore, not all the molecules can end up on a single high energy level, for the total energy would then be more than that available. Nor, in general, can they all end up on the lowest energy level, because then their total energy would not match the energy available. ('In general' is always a weasel phrase but is intended to imply that there might be special cases that allow exceptions from a general rule: I shall return to the point in a couple of paragraphs; live with it for now.) Boltzmann's derivation of his distribution took this constraint into account, and the distribution I have described, with successively fewer molecules in levels of increasing energy, is the actual outcome. In short, the Boltzmann distribution is the outcome of anarchy in alliance with indolence: the nearly random populating of energy levels, anarchic behaviour, subject to the conservation of energy, that outcome of indolence.

This is where I need to introduce another word picture, this time the form of Boltzmann's actual expression for his distribution. It turns out that the gradual lessening of populations with increasing energy of the levels is described by a very simple mathematical expression.[1] Moreover, that expression depends on the value of a single parameter. When that parameter has a small value, the populations fall off very quickly with increasing energy and only the bottom few energy levels are populated (but still with the typical falling off as levels of higher energy are considered). When the parameter has a high value, the populations reach up to high energy levels: although there will be most molecules in the lowest energy level, fewer in the next higher, and so on, there will now be molecules present with very high energies. The expression and the parameter in question are 'universal' in the sense that they apply to any kind of substance and any kind of motion. That is, for a given value of the parameter, the relative population of a level of a given energy is the same, regardless of whether it refers to vibrations or rotations of molecules or of whether it refers to the vibrations of atoms in solids, or if the substance involved is lead or lithium, chalk or cheese, or any other substance.

The name given to this universal population-controlling parameter is 'temperature'. I hope you now have some insight into its nature. A low temperature describes a Boltzmann distribution in which only low energy levels are occupied and with diminishing populations on stepping up the levels to higher energies. A high temperature describes a Boltzmann distribution in which the populations spread up into high energy levels, and the higher the temperature the greater the reach.

Before leaving this point, I need to scotch that 'in general' of a couple of paragraphs ago. Suppose the value of the parameter, the temperature, is set equal to zero. In this case, according to the form of the Boltzmann distribution with this value of the temperature, all the molecules will be in the lowest energy level; there are no molecules at all in any other energy level. All the books are on the bottom shelf. This is the 'absolute zero' of temperature, with no further lowering of the temperature physically meaningful, for how can the molecules occupy an even lower energy level than the lowest? Of course, this particular distribution still has to respect the conservation of energy, so it is attainable only when all the energy has been sucked out of the sample and the total energy is effectively zero. ('Effectively' is another very useful weasel word; I shall let it slip by. I have put it there, simply as an unreconstructed pedant, to let other pedants know that I know what they are, or should be, thinking.[2])

* * *

So much, for now at least, for the molecular interpretation of temperature and the insight that Boltzmann's distribution brings to its meaning. The measurement of temperature was well established, but the concept obscure, well before Boltzmann had hanged himself, and the familiar everyday scales (Fahrenheit's and Celsius's in particular) had long been established in a pragmatic way with each inventor choosing readily reproducible and transportable 'fixed points' to set the scale. Thus, Daniel Fahrenheit (1686–1736) set the zero of his scale at the lowest attainable temperature then readily attainable (and well above the absolute zero that I have been discussing), specifically the freezing point of a mixture of common salt and water, and he took as 96 (puzzlingly, not 100) the temperature of his readily transportable armpit, or at least of the average ubiquitous armpit. The 96 gradations between these two vaguely fixed points then result in the freezing point of pure water turning out to lie at 32 on his scale and the boiling point of water at 212, well above the temperature of his armpit. Anders Celsius (1701–44) rather more wisely focused on the properties of water itself to specify the fixed points, setting 100 at its freezing point and 0 at its boiling point. His scale has since been inverted (I explore the wisdom of that in Chapter 9), so that hotter things have higher temperatures than cooler things. It is a minor point of interest, that according to their definitions both are 'centigrade' scales in the sense of both having about 100 gradations, degrees, between their fixed points; but modern society, seeing 32 and 212 rather than Fahrenheit's original underlying 0 (his salt mixture) and 96 (his armpit), thinks only of the Celsius scale as synonymously centigrade.

Just to round out this mention of temperature scales, the scale that sensibly sets absolute zero at 0 is called the 'thermodynamic temperature scale' or more colloquially the 'absolute temperature'. If the gradations of this scale have the same size as Celsius's scale, then it is known as the 'Kelvin scale', named after William Thomson, Baron Kelvin of Largs (1824–1907), a pioneer of thermodynamics.[3] If the gradations are the same size as Fahrenheit's, then the thermodynamic temperature scale is known as the 'Rankine scale', named for the Scottish engineer John Rankine (1820–72), now but not then a lesser-known theoretician of steam engines and composer of comic ditties. Hardly anyone, as far as I know, uses the Rankine scale any more: perhaps engineers still do in America, where in everyday life Fahrenheit stubbornly refuses to cede victory to Celsius. For completeness, absolute zero lies at −273.15 °C or −459.67 °F.

After that excursion into the pragmatic, the question I now need to explore is how the concept and significance of temperature entered science, and specifically thermodynamics, as an observable property before there was acceptance of the reality of molecules and no anticipation that their energy levels were discrete. That is, what was temperature before Boltzmann?

Temperature entered thermodynamics formally as an afterthought. You need to know that a feature of thermodynamics is that each of its laws typically (there's that weasel word again) introduces a new property relating to energy. Thus, the first law of thermodynamics introduces the property we know as energy itself; the second law of thermodynamics (to make its entry in Chapter 5) introduces the property we know as 'entropy'. Both those laws make use of the concept of temperature in a variety of ways, and it slowly dawned on the originators of thermodynamics that although they had rigorous statements of the first and second laws, and therefore definitions of energy and entropy, temperature itself hadn't been defined by the statement of a law. A new, more fundamental law than either the first or the second had to be formulated, a law that formalized the definition of temperature. At that point, with 1 and 2 used up, the founders of thermodynamics simply had to grit their teeth and resort to calling their new law, preceding logically as it did the first and second laws, the 'zeroth law of thermodynamics'. (I am not aware of any other branch of science that has needed to introduce a zeroth law as an afterthought: perhaps there is one lurking unstated in Newtonian classical mechanics.) In short, the zeroth law introduces temperature in a formal way, and I need to explain its seemingly rather banal content and how it does its job.

Suppose you have three objects which I shall call A (for instance, a block of iron), B (a bucket of water), and T (if you were expecting C, hold on). One characteristic that will now become clear is that there is something rather odd about thermodynamicists, the practitioners of thermodynamics: they get really excited when they notice that nothing happens. You might have noticed that in the discussion of the conservation of energy in Chapter 2: they really drooled (in their abstract way) when they noticed that the total energy of the universe didn't change. The drooling resulted in the first law of thermodynamics, which is a kind of elaboration of the law of the conservation of energy. Here is another, for them exciting, scenario. Suppose you put A and T into contact and notice that nothing happens. Now suppose that, separately, you put B and T in contact and nothing happens. The zeroth law says that *if you now put A and B in contact with each other* (put the block of iron into the bucket of water), *then nothing will happen*. This is a universal observation: whatever the nature of A and B, if nothing happens when each in turn is put in contact with T, then nothing will happen when A is put in contact with B. That observation, for a thermodynamicist, is almost overwhelmingly orgasmic and leaves them suffused with joy.

I hope you can now see that object T is playing the role of a thermometer, hence the T, and in some sense is measuring temperature. Thus, when A is put into contact with T and nothing happens (such as a thread of mercury remaining the same length inside T's glass tube), it means that A has the temperature represented by the length of the thread of mercury. When B is put in contact with T and nothing happens, it means that B has the temperature recorded by T, the same as that of A. So, A and B 'have the same temperature' and we can be confident that when put in contact with each other, nothing will happen. This cycle of nothing happenings is how the zeroth law introduces the concept of temperature.

Now I need to link the concept of temperature as it is introduced by the zeroth law to its molecular interpretation in terms of Boltzmann's distribution. The key point, as I stressed earlier, is that temperature is the parameter that characterizes the distribution of molecules over the available energy levels and is universal (specifically, independent of the substance being discussed). Object A (the iron) has an array of energy levels and its atoms spread over them in accord with the Boltzmann distribution for the prevailing temperature. Object B (the water) has an array of energy levels and its molecules spread over them in accord with the same Boltzmann distribution (the parameter, the temperature, is the same as for A). When A and B are brought together (the iron is immersed in the water), their energy levels interlace like the fingers of your two hands when they, your hands, are brought together. The distribution of molecules remains the same, so the temperature remains the same, and, in short, nothing happens.

The Boltzmann distribution captures and, like all molecular interpretations of phenomena, enriches the concept of temperature. You can now begin to see why it also captures the stability of matter in the everyday world and the ability of matter to undergo change when heated. At normal temperatures, the distribution of populations doesn't spread up very far in energy and most molecules are in low energy levels, able to do little except, for instance, vibrate languidly. Thus matter is long-lived. When the temperature is raised, more and more molecules occupy their high energy levels, and as for life so for molecules, high energy means things can get done. In particular, atoms can be thrown off and new bonds formed: chemical reaction can occur. In the kitchen, cooking is the process of using an oven or hob to push molecules higher up their available energy levels until sufficiently large numbers of them have enough energy to react. Refrigeration brings molecules down into their lowest energy levels and calms them. We would say, preserves them.

There is a law in chemistry about the rates at which reactions take place and which springs from the Boltzmann distribution. The Swedish chemist Svante Arrhenius (1859–1927), who received one of the earliest Nobel Prizes (in 1903), which he himself had helped to establish, proposed what is now known as the 'Arrhenius rate law', that the rate of a chemical reaction increases with temperature in a specific way that depends on a single parameter known as the 'energy of activation', which varies from reaction to reaction.[4] Broadly speaking (meaning that there are many exceptions), the rate of a chemical reaction typically doubles for each 10 degrees rise in temperature. The explanation lies in the Boltzmann distribution, for the activation energy is simply the minimum energy that molecules must have in order to react, and the number that can do so rises with temperature as the Boltzmann distribution stretches up into higher levels. Cooling (refrigeration) has the opposite effect: as the Boltzmann distribution sinks into lower levels, fewer molecules have the energy to react and the reaction slows.

Arrhenius's law has numerous everyday consequences. We cook food by raising molecules to levels above their activation energy by raising the temperature by many tens of degrees and thereby speed the reactions that break down foodstuffs. We preserve food by shrinking the Boltzmann distribution so that few molecules have the energy to react. The body fights disease with fever, by raising its temperature to upset the delicate balance of the rates of chemical reactions that keep us, and the invading bacteria, alive (there is clearly the need for a delicate balance here!). Fireflies flash faster on warm nights than on cold. Industry deploys the rate law to bring about the reactions it needs to spin materials from the feedstock. All around us is a chorus of chemical reactions buzzing away in accord with the Arrhenius law and displaying with their changing rates the modification of the Boltzmann distribution, with its foundation in anarchy, which changes in temperature bring about.

* * *

The question perhaps forming in your mind is what happens when two objects that do not have the same temperature are brought together and placed in contact. Now we leave the realm of the zeroth law and the excitement of nothing happening and move into the domain of laws of nothing not happening. But before going there in any formal sense, you already know from everyday experience what happens: energy flows from the hotter body into the cooler body (the hot iron cools and the cool water warms), and in due course the two bodies reach an intermediate temperature and resume their life of 'thermal equilibrium', with nothing apparently happening. I shall develop this familiar feature of Nature and use it to introduce another law that has wide ramifications and, needless to say, also springs from anarchy.

I need to introduce you to the notion of 'heat'. In a sense, that is easy, because there is no such thing. Despite its usage in common conversation, a hot body does not contain heat; it does not lose heat as it cools, for it had none to begin with. Despite what many say, and possibly what you already think, heat is not a form of energy. In science, heat is not a thing, it is a process. Heat is *energy in transit as a result of a difference of temperature*. Heating is not the process of increasing the heat of an object; it is a process that can be used to increase its temperature (doing work—stirring a liquid vigorously—can also raise its temperature). Heating, if not to the cook but to the scientist, is the process of transferring energy to an object by making use of a temperature difference; they are not transferring something called 'heat'. Heat was indeed once thought to be a fluid, then known as *caloric* (from the Latin *calor*, meaning heat, going far back with its fascinating etymology to the Sanskrit *carad*, meaning harvest, the 'hot time'), and its flow has many of the attributes of a fluid; but that was early in the nineteenth century, and the interpretation has been overthrown. All these fussy remarks undermine the everyday meaning of 'heat', but science typically proceeds by adopting an everyday term and refining its meaning. In this case the everyday unrefined term 'heat' which as a noun seems to imply its possession as a property ('this furnace gives off a lot of heat') is distilled into the fine spirit of process, a process in which energy is exchanged as a result of a difference in temperature. Pedantry can be either a cleansing of the brain or its stuffing up; I hope the former in this case, but it is important to be clear about the meaning of words, nowhere more so than in science where truth depends on precision. Chapter 9, on the role of mathematics in expressing the laws of nature, takes this attitude to precision of expression to its extreme, and shows how it underlies the extraordinary power of mathematics in science. So, let's move on with the meaning of heat in mind.

First, we need to consider the interpretation of transferring energy as heat in terms of the Boltzmann distribution. Let A (the iron) and B (the water) now have different temperatures, with A hotter (that is, having a higher temperature) than B. You know what that means observationally: when each is put in touch with different bodies T ('different thermometers'), nothing happens provided the length of the mercury thread in each T has a different length corresponding to the temperature of either A or B. Now look underneath the surface into the underworld of the atoms making up A and B. Because the parameter we are calling 'temperature' is different in each case, the atoms of A have a different Boltzmann distribution over their energy levels than the molecules of B have over theirs: in hot A the atoms populate levels higher in energy than those of cooler B.

When the two bodies are brought into contact all the energy levels of A and B are available to all the atoms (as before, think of your fingers, representing the two sets of energy levels, as interlaced, or a single bookcase formed by merging two). Once equilibrium has been restored there is a single Boltzmann distribution of the atoms over this single set of levels. To achieve that distribution, some atoms must tumble down from the high energy levels provided by A into the lower ones, which are either A or B, until the populations match Boltzmann's most probable distribution. As a result of this tumbling, the levels of once cool B will become more heavily populated at the expense of the populations in once hot A, including some of the higher levels of B that were very sparsely populated initially. Thus, the joint system becomes characterized by a single temperature with a value intermediate between the two initial values. The iron has cooled and the water warmed.

Here is a slightly pedantic point (I cannot resist them). When you leave a cup of hot coffee to stand, it cools to the temperature of its surroundings, not to some intermediate value. What is going on in Boltzmann's eyes? Even the hot block of iron you have had in mind so far would simply cool with no discernible effect on its surroundings. The key word here is *discernible*. The explanation is that the surroundings (the table, the room, the Earth, the universe...)

are so vast that to accommodate the incoming energy their myriad energy levels undergo an almost completely negligible redistribution of populations. In other words, although energy has been accommodated and a tiny redistribution of populations has occurred, the redistribution is unnoticeable and, therefore, the temperature of the surroundings is for all practical purposes unchanged. It is rather like noting that a vast piece of blotting paper remains white even though it has soaked up a droplet of ink.

What I haven't yet mentioned is time, and in particular the time it takes for an object to cool to the temperature of its surroundings. This switch of focus from temperature to time introduces the law I am aiming towards and its important ramifications. Anarchy will still be the origin. To show its role I need first to outline the regularity of the phenomenon, the law of cooling, and then move on to describe the underlying anarchy as atoms go about their business oblivious of any rules.

The rate at which a body cools to the temperature of its surroundings is summarized by *Newton's law of cooling* (the usual Newton), which he published, apparently anonymously, in 1701. The law summarizes his and since then countless other observations on cooling by stating that *the rate of change of temperature of a hot object is proportional to the difference in temperature between it and its surroundings.*[5] A very hot object (compared to its surroundings) cools rapidly initially, then as its temperature falls the rate at which its temperature falls declines, and vanishes entirely when it has reached the same temperature as its surroundings. This type of behaviour, in which the value of a property falls at a rate proportional to the current value of the property (in this case the 'property' is the temperature difference) is called an 'exponential decay' of the property. The term 'exponential' is widely misused in common discourse, often being taken, as in 'an exponential increase in population', to be something like 'strikingly or worryingly big'. I shall use it in its precise sense, as described here (as current change proportional to current value). You should be aware that an exponential change could be extremely slow, as in the exponential but almost undiscernible cooling that occurs when the temperature of an object is almost the same as its surroundings. Lest it appear that I am making too big a meal of cooling, I would like to remark that the exponential cooling characteristic of Newton's law has its analogues throughout science, in topics far removed from cooling, and I shall stand on Newton's cooling shoulders to introduce another of them shortly.

The first important feature that I need to introduce and have ignored so far is that the molecules in a Boltzmann distribution are not just lying on their shelf-like energy levels: they are ceaselessly jiggling between the available levels. Just as a Dickens might suddenly tumble on to a lower shelf and a Trollope take its place by being boosted up from a lower shelf, with overall the distribution remaining Boltzmann-like, so molecules are ceaselessly migrating between their levels. In the underworld of atoms, all is motion, migration, and re-accommodation. That is, and this is a very important point, the Boltzmann distribution is a dynamic, living beast, pulsing with internal change. It is the most probable distribution of a ceaselessly changing, fluid, dynamic underworld. The calm observed by an external observer conceals the storm within.

The second feature I need to introduce is the rate at which an individual molecule jumps between levels as a result of all the jostling going on. This rate can vary widely, with some molecules lingering for ages in a level, but then rattling rapidly between many of them. You should think of each molecule as occupying a given energy level for various lengths of time, with an average lifetime of a tiny fraction of a second, before moving on. The principal point is that the behaviour of an individual molecule (specifically its lifetime in a given state) is wholly independent of what other molecules are doing: each molecule is an island.

Now imagine bringing two objects (iron A and water B) together. The redistribution of molecules takes place as I have described, but now we need to blend into the discussion the feature that individual molecules are migrating at the same constant, average rate. The average number that make a jump to another level in a given interval depends both on the average lifetime (the shorter that average, the more will have jumped at the end of the interval) and also on the number that are poised to jump (the greater that number, the more will have jumped during that interval). Therefore, the rate at which a population of a level jumps to other levels depends on the average lifetime (the shorter the faster) and the population (the faster the population changes the more molecules there are poised to jump). Here is the crucial point. A lot of molecules in high levels are lining up to become redistributed when A is much hotter than B, so the redistribution will then be fast. When the temperatures are nearly the same, there will be only a small number of molecules that need to undergo redistribution, so that redistribution will be slow. In short, the rate of redistribution is proportional to the difference in distributions. Bearing in mind that the distributions depend on the temperatures, the rate of change of temperature is proportional to the difference in temperature between the two objects. This proportionality implies that the cooling is exponential, which is the content of Newton's law of cooling.

The crucial point is that if the molecules are allowed to jump between levels without constraint, the outcome is a law, that of exponential decay. Once again, anarchy has resulted in a law. Exponential decays (and in some cases, exponential rises) are common in physics and chemistry, and all of them stem from anarchic underlying behaviour in which individuals undergo change randomly and independently of what other individuals happen to be doing.

One important example is the *law of radioactive decay*, in which the activity of a radioactive isotope decays exponentially with time.[6] Radioactivity stems from the fragmentation of a nucleus (for instance, the nucleus tossing off an alpha particle or a beta particle) or undergoing internal collapse and generating a gamma-ray photon (or a combination of these processes), with each nucleus having a constant probability of fragmentation in a given period. These processes occur independently of what a neighbour nucleus happens to be doing and so give rise to exponential decay.

For instance, a carbon-14 nucleus (a nucleus of carbon with six protons and eight neutrons instead of the usual six neutrons) has a certain probability of emitting a beta particle, a fast-moving electron, each second (the probability is known to be one in 250 billion, so you have to wait a long time before you can be confident that a given nucleus will eject a beta particle). This individual probability of decay is the same for all the carbon-14 nuclei in a sample and is independent of the external conditions and what is happening at the nucleus next door. It depends only on the details of how the components of the nucleus are bound together by the forces acting between these components. Once a beta particle has been emitted, the nucleus (in carbon-14's case, it becomes nitrogen-14 with seven protons and seven neutrons) becomes inactive and ceases to emit rays. The overall rate, however, of all the nuclei in the sample emitting beta particles therefore declines with time, with the number decaying in any given period proportional to the number available to decay. Initially the overall rate is high, but as nuclei die, so the rate declines, exactly like the temperature difference in Newton's case, and the rate of radioactive disintegration decays exponentially.

There are important consequences of the law of radioactive decay. A positive consequence is the ability to use the law to assess the age of organic artefacts by 'carbon-14 dating', in which the relative abundance of carbon-14 and carbon-12 (the common, stable isotope) changes exponentially with time. Less benign is the slow decay of many radioactive isotopes, especially those left over from nuclear fission processes in power stations and explosions. A mathematical consequence of exponential decay is that it takes the same time for an isotope to decay to half its initial abundance, then to half that again, to half that again, and so on. That time is the 'half-life' of the isotope (for carbon-14 it is an archeologically handy 5730 years). Although intensely radioactive isotopes might have half-lives of fractions of seconds, some are measured in years and even thousands of years. There is nothing we can do to change that, apart from changing the identity of the isotope by yet another nuclear process and turning it into one that is short-lived.

* * *

This chapter has covered a lot of ground, and once again a synopsis might be helpful. I argued that the overwhelmingly most probable outcome of the random distribution of molecules over their available energy levels (subject to the constraint of the conservation of energy) is the Boltzmann distribution, a dynamic spread of populations that depends on a single universal parameter, the temperature. This distribution behaves as you would expect for the common notion of temperature and helps to explain why matter is stable under normal conditions, but as the temperature is raised increasingly becomes able to undergo transformation into different substances. I also pointed out that if no constraint is put on the behaviour of individual, independent molecules, but they are allowed to go about their business in a random way, then we end up with a type of behaviour that occurs widely in Nature, namely exponential decay. That discussion led to an explanation of two laws of nature, Newton's law of cooling and the law of radioactive decay.

Indolence and anarchy have raised their heads above various parapets throughout this discussion. They (through quantum mechanics) account for the existence of energy levels. The details of the derivation of the Boltzmann distribution have depended on the indolence-based conservation of energy and the anarchic, random distribution of molecules over their energy levels. The rates at which temperatures equalize, a representative of various types of change, is also based on the individual anarchy of the behaviour of individuals, unconsciously conspiring to generate, or perhaps just stumbling into, a law.

5

Beyond Anarchy

Why anything happens

I have mentioned, in Chapter 4, that thermodynamicists, the folk who study and apply thermodynamics, become very excited when nothing happens. If to their disappointment something does happen, then they take pleasure in noting that things invariably get worse. That, the observation that things get worse, is the *second law of thermodynamics*, one of my favourite laws of nature. Of course, in science this populist statement of the law is dressed up in formal clothes and given power by expressing the same observation more precisely and mathematically, but 'things gets worse' is its essence. As another prefatory remark, I have also mentioned that each law of thermodynamics introduces a new property relating to energy and aspects of its transformation: that afterthought, the zeroth law, introduced temperature and the first law introduced energy. The second law introduces a third major property, the 'entropy'. My aim here is to show that the second law is yet another manifestation of indolence and anarchy, but that the properties of entropy account for the emergence of structures, events, and opinions that might be exquisite.

Things get worse. I need to elaborate that remark and develop it to the point that shows how it enables, and actually causes, the emergence of the exquisite. The remark about getting worse is of course a faintly jocular interpretation of the formal statement of the second law, which states that *in a spontaneous process, the entropy of an isolated system tends to increase.* There are several terms in this more austere statement that I need to explain, but don't let them cloud your overall impression of the content of the law, that the universe is drifting unstoppably into worseness.

I need to explain, I hope without too much turgidity, 'spontaneous process', 'entropy' of course, and 'isolated system'. A 'spontaneous process' is an event that can take place without external intervention, without being driven; it is a natural change, like water running downhill or gas expanding into a vacuum. Spontaneous does not mean fast: some processes might be spontaneous but take ages, even eons, to unfold, like treacle running downhill or the motion of glaciers. Other spontaneous processes may be over in a twinkling, like the expansion of a gas into a vacuum. Spontaneity, in this context, is all about tendency, not the speed of realization of that tendency.

The word 'entropy' is from the Greek for 'turning towards' and was coined in 1856 by the German physicist Rudolf Clausius (1822–88), who figures again in this account below. Entropy is a measure, one that can be defined precisely, of disorder: simply put, the greater the disorder, the greater is the entropy. Several people came up with the actual quantitative definition of entropy: among them was Boltzmann, the hero of Chapter 4. His formula for expressing entropy as a measure of disorder, not to be confused with the formula for his distribution, is carved on his tombstone in Vienna.[1] We don't need to know it: I deal here with interpretations not equations. 'Increased disorder' is often easy to identify, but sometimes it is dressed in subtle deceptions. I shall give examples later.

Finally, 'isolated system' means the part of the world we might be interested in ('the system') but cut off from all interaction with its surroundings. No energy can leave or enter an isolated system, nor can matter. Think of stuff inside an opaque (to prevent radiation entering or leaving), rigid (to prevent energy being used to do work of expansion), sealed (to prevent matter entering or leaving) vacuum flask (to prevent energy entering or leaving as heat). When applying the second law, the concept of an isolated system plays a crucial role. If you want to think big, then the entire universe is an isolated system (or so we take it to be). You should be aware that a thermodynamicist might, with some modesty, think small and regard a stoppered flask immersed in a water bath as their entire universe.

As I develop this account of the second law and show how the loss of form can generate form, how it underlies the majestic march of evolution, and how it accounts for the emergence of the horrid as well as the exquisite, I need you to accept that there is a natural tendency for matter and energy to spread in disorder. There are deep questions here that I shall need to return to, but for the time being I hope you can take it as 'obvious' that if atoms and molecules are able to wander around at will—the 'at will' meaning that they are not directed to travel in a particular direction or to assemble in a particular way—then it is far more probable that a structure will decay into disorder than it is for disorder to assemble into structure. Thus, the molecules of a gas injected into the corner of a container are far more likely to spread out to fill the container than for the molecules of a gas that fills a container uniformly to cluster, without external intervention, into one corner. Of course, you could squash them up into one corner by using some kind of piston arrangement, but that would be external intervention, which is disallowed in an isolated system. Similarly, the energy of vigorously vibrating atoms in a hot block of iron are far more likely to jostle their neighbours in the surroundings and hence disperse their energy among the latter's molecules than for energy to accumulate in the block by random jostling of the external molecules and for the block to become hot at the expense of the cooler surroundings. Once again, you could contrive a way of heating the block by using energy from the surroundings, but contrivances are external interventions, and are disallowed in an isolated system.

* * *

The direction of natural change is for matter and energy to disperse in disorder, unhindered by rules except the overriding indolence-originating one of the conservation of energy. This remark can be expressed in a different way: although the *quantity* of energy in the universe remains constant, its *quality* tends to degrade. Energy that is concentrated in one location is high quality in the sense that it can be used to do all manner of things (think of a litre of fuel); once released and dispersed (such as by combustion), that energy is still somewhere but is now far less useful. A high-pressure gas in a cylinder represents localized, high-quality energy as its molecules speed round in the confined space. That quality is degraded if the gas is allowed to escape and the energy of its molecules dispersed. Here is thermodynamics in a nutshell: the quantity of energy survives; its quality declines.

Entropy is simply a measure of that quality, with high entropy meaning low quality. The entropy of a fuel is low; that of its combustion products is high. The entropy of a compressed gas is low; after it has expanded its entropy is high. So 'quantity survives, quality declines' becomes 'energy survives, entropy increases'. Likewise, the jocular 'things get worse' becomes more formally 'entropy tends to increase'.

When entropy was first introduced into the world of science, in the 1850s, there was considerable puzzlement about where it came from. The Victorians were comfortable with the constancy of energy, for (in their view) it didn't have to come from anywhere once the Creator had sufficiently endowed the universe with what He judged in His infinite wisdom to be exactly enough to serve our needs forever. But entropy seemed to be coming from nowhere. Was Creation still going on? Was there a deep undiscovered inexhaustible well of entropy slowly being pumped out into our perception at a rate judged appropriate by that same infinitely wise Creator? Science came to the rescue of this culturally appropriate but simplistic view of the nature of reality, as in so many other instances, in the form of a molecular understanding of entropy.

* * *

Whenever a change takes place, the disorder of the universe increases, the quality of its energy degrades, its entropy increases. The funny thing is, such is the interconnectedness of events in the world, that this degradation is not a cosmically uniform sliding into disorder, a general elimination of structure, a global dispersal of energy, a collapse of matter into slime. There may emerge local abatements of chaos, we among them. The only requirement of the second law is that the total entropy of an isolated system (the universe, or an isolated part of it such as that little water bath and its flask) increases in a spontaneous change: in localized pockets the entropy may decrease and a structure emerge provided that overall there is an increase in disorder.

Let's see in more detail what this means. Consider an internal combustion engine. The fuel is a compact concentration of stored energy. When it burns, its molecules fragment (if it is a hydrocarbon, they become lots of little carbon dioxide and water molecules) and disperse. The energy released in the combustion spreads into the surroundings. The arrangement of pistons and gears in the engine is designed to respond to this dispersal and spreading and in effect to capture it. The engine might be part of a crane being used to build a cathedral, and being used to lift blocks into position. Thus, as dispersal occurs in the engine itself, a structure emerges elsewhere. Overall the universe has become more disordered, but locally, at the cathedral, a structure has emerged. Overall there is an increase in disorder; locally there has been an abatement.

You can find this production of order driven by collapse into greater disorder wherever you look. There are often linked chains of one increase in disorder driving order somewhere which then crumbles and in its crumbling generates order elsewhere. All that matters in this concatenation of events is that more disorder emerges somewhere than the disorder it destroys elsewhere. That destruction of disorder is the generation of order.

The Sun is the great disperser in the sky and through the concatenation of dispersals drives events, including evolution, on Earth. The nuclear fusion events that occur within the Sun release energy that disperses through space. A tiny fraction of that energy is captured by green vegetation on Earth and is used to build organic structure. In this case, the disordered starting materials are carbon dioxide and water, and the highly organized structures that become generated are the carbohydrates that cover harsh lithosphere with benevolent biosphere. Organic structures: plants, trees, you name it, have been driven into existence, but the Sun has died a little and the solar system is more disordered despite the growth of vegetation on Earth.

That vegetation is food for animals. Food is fuel for the internal combustion that powers us and them. The combustion of food is much more subtle than the combustion of a fuel in an engine—there are no flames within us—but it is analogous in so far as digestion degrades complex molecules into little molecules, including carbon dioxide and water, and energy is released. Organisms are not arrangements of gears and cogs, but the metabolic processes inside them (and us) are the organic analogues of cogs and transmit the organizing power of the initial digestion to the sites analogous to cranes. There, amino acids, some of the little molecular building blocks of Nature, are hoisted into the structures we know as proteins, intricate little molecular cathedrals, and the organism grows. Overall, noting the digestion of the meal, there has been an increase in disorder, but the organism's biochemistry has tapped engine-like into that dispersal and a structure, perhaps you, has emerged. As I have indicated, we are local abatements of disorder; we are the children of chaos.

It is not only you and I who are driven into existence by the generation of increased disorder. The entire ecosystem is a descendant of disorder and a consequence of chaos. No organism is an island. Natural selection is Nature's extraordinarily complex and wonderful way of accommodating to the second law. The biosphere is an extraordinary reticulation of interdependent entities, one feeding off the other, literally to live off the dispersal the devouring develops. Fuel, the eaten, is scarce and essential to survival and later propagation, for life is structure and must be sustained by increasing the disorder of the universe. Living things simply cannot have avoided living off each other, and natural selection resulting in evolution is a consequence.

There are still those in the world who cannot reconcile the collapse into disorder that the second law identifies as the spring driving change with the emergence of the elaborately organized structures known as organisms. They cannot see that dispersal engenders structure. The resolution of this difficulty is the point I have stressed several times already. The only necessity is that the *overall* disorder is increasing. Linked as one event might be to another, a local patch of increasing disorder (the consumption of fuel, the eating of an antelope, and myriad other possibilities, including that civilized elaboration of disorder-generation we know as a dinner party) can drive another patch of the universe from disorder into order. All that is necessary, apart from the mechanism linking the patches, is that the increase in disorder outweighs the decrease in disorder so that overall there is a rise in disorder. Countless processes are examples of this interplay embedded in the second law; evolution by natural selection is merely the most exhilarating.

* * *

I have focused mostly on organisms, for that is where the second law illuminates brilliantly and perhaps surprisingly. There are plenty of other purely inorganic and technological manifestations of the law. Most of the applications of the second law in technology stem not from Boltzmann's tombstone formula for entropy but from an alternative expression proposed by Clausius in 1850. Ignorant of the molecular interpretation of entropy, he proposed what at first sight seems to be a wholly unrelated expression for the change in entropy accompanying a process in terms of observable properties (as distinct from the disorderly dispersal of energy and molecules). He proposed that the change in entropy should be calculated by monitoring how much energy was transferred as heat into or out of the system and dividing the result by the temperature at which the transfer has taken place.[2]

Clausius didn't associate the outcome of his calculation with disorder, but we can. Transfer of energy as heat makes use of the random jostling of neighbouring molecules, such as those in a hot flame or the vigorously shaking atoms in an electric heater. That jostling stirs up the molecules in the system of interest into disorderly motion and thereby increases the entropy. So far so good, for transferring energy as heat increases the entropy. But why does the temperature matter? The analogy I like to use is that it is like sneezing in either a busy street or a quiet library. A busy street is the analogue of a hot object with a lot of thermal turmoil. A quiet library is the analogue of a cold object, with atoms not jiggling around very much. The sneeze is the analogue of an injection of energy as heat. When you sneeze in a busy street, the increase in disorder is relatively small. When you sneeze in a quiet library, the increase in disorder is substantial. So it is with Clausius's definition: energy transferred as heat to a hot object doesn't increase the disorder very much, so the change in entropy is quite small. When the same amount of energy is added as heat to a cool object, the change in entropy is large. The temperature in Clausius's formula captures the difference between street and library.

Clausius's approach also captures a very important result established in the early days of thermodynamics by the French engineer Sadi Carnot (1796–1832) in work that lay largely ignored for decades as his conclusions were so outlandish and not consistent with the common-sense attitudes of the engineers of the day.

He argued (using concepts we would now regard as incorrect, for instance treating heat as 'caloric', an imponderable (weightless) fluid, that water-wheel-like generated work as it trickled through an engine) that the efficiency of an ideal steam engine depends only on the temperatures of the hot source from which energy is drawn as heat and the temperature of the cold sink into which energy is discharged.[3] He showed, perhaps more remarkably, that the efficiency is independent of the identity and pressure of the working substance (typically steam).

Carnot's result is not a new law of nature, but it illustrates how a law, in this case the second law of thermodynamics, can spread its net and capture all kinds of behaviour. Here is the argument. Imagine an engine as consisting of a hot source of energy, a cool sink into which energy can be discarded, and between them a device for using energy to do work (think of it as some kind of turbine). Now imagine drawing some energy as heat out of the hot source. The entropy of the source goes down, but because the temperature is high, Clausius's formula implies that the decrease is not great (the hot source is like a busy street). The energy you have extracted is turned into work by using some kind of mechanical device. At this point you should be able to see that not all of that energy can be turned into work. If it were, there would be no further change in entropy and overall the entropy would go down. That means the engine wouldn't work because for a natural change to occur, entropy must increase.

In order for the engine to work, some of the energy you drew from the hot source must be discarded into the cold sink (which might be the atmosphere or a river). That injection of energy as heat into the cold sink will increase its entropy. Even a small amount of energy injected will have a big effect on the entropy of the sink because the temperature of the sink is low (it is like a quiet library). But how much energy must you discard in this way and be unable to use it for doing work?

The engine can do most work if you discard the least possible amount of energy. That least amount must be enough to increase the entropy of the cold sink just enough to overcome the reduction due to drawing energy from the hot source. Because the cold sink is at a low temperature, a large increase in entropy is obtained by discarding even a small amount of energy as heat into it. The precise amount depends on the temperature of the two sources and nothing else. The consequence is that the efficiency of the engine, which depends on how much has to be discarded relative to the amount extracted, depends only on the two temperatures and is independent of any other details about the construction and operation of the engine. To achieve the greatest efficiency you need a source that is as hot as possible (so that the reduction in entropy of the hot source is minimal) and a sink that is as cold as possible (so that even a tiny 'waste' of energy generates a lot of entropy). That is what Carnot had concluded early in the nineteenth century to the sceptical response of his audience. But he was right.

There are several alternative, equivalent statements of the second law of thermodynamics which do not mention entropy, but which you can now understand in the light that entropy throws on them, and this little discussion of Carnot's work. Both statements I intend to present illustrate one of my favourite quotations, originally made by the Hungarian biochemist Albert Szent-Györgyi (1893–1966), who said words to the effect that being a scientist involves seeing what everyone else has seen but thinking what no one else has thought. William Thomson (Lord Kelvin of Largs, 1824–1907; he took his title, in 1892, from the River Kelvin, which runs near his laboratory in Glasgow) saw, like many before him, that *a steam engine wouldn't work unless it had a cold sink.* He thought about that, and used it as the basis of the 'Kelvin statement' of the second law of thermodynamics and spun a web of thermodynamics from it.[4] We now know why: if there is no cold sink, there can be no increase in entropy and so such an engine would be impotent. Rudolf Clausius was also a scientist who saw but then thought. He noticed (I fantasize anachronistically here) that *for it to work, a refrigerator had to be plugged in.* More precisely, he noted, as everyone knew but no one had thought about, that *heat did not flow from a cool body to a hotter body without doing work to bring it about.* He developed this observation into what we now call the 'Clausius statement' of the second law and spun his version of thermodynamics from it.[5] We now know why: if energy leaves a cold object as heat, there is a big decrease in entropy, and when it enters the hot object there is only a small increase. Overall, there is a decrease in entropy, and the process isn't spontaneous. Work has to be done to bring it about: the refrigerator must be plugged in. The flow of energy from the cold object must be augmented by doing work, so that when it enters the hot object it can stir up enough entropy to overcome the lowering of entropy of the cold object. One feature to note is how two apparently disparate statements of a law of nature, Kelvin's and Clausius's, are melded into a single statement by introducing what originally was thought to be an abstract statement in terms of entropy. Abstraction is an extraordinarily powerful tool for consolidating the seemingly disparate, forging progress, and developing insight.

Here is another perhaps surprising insight from this analysis, yet another topsy-turvying of common sense. An argument can be made that the most important component of an engine is its natural surroundings, the atmosphere or a river, not its heavily engineered components. As you have seen, what drives change is an increase in entropy, which in an engine is achieved by transferring energy as heat into a cool sink. If that increase doesn't occur, the engine is impotent, so the essential component is where entropy increases, the natural surroundings. I accept that to achieve that increase there must be a supply of energy as heat, which comes from the hot source and after working the turbine or piston is dumped into the sink. But that supply, in a sense, is secondary, and in fact works against what you are trying to achieve, for the withdrawal of energy as heat from the hot source contributes a small lowering of entropy, which is the opposite of helpful for driving the engine. The potency of an engine is really an aspect of its surroundings, with the hot source a secondary but necessary evil.

Engineers use Carnot's conclusion in their quest for improving the efficiencies of engines and a range of related machinery, such as refrigerators and heat pumps. These technological applications all depend on the ultimate vision of Nature that I have presented, in which the universe gradually sinks into disorder with no guiding principle except deep underlying indolence-based laws such as the conservation of energy.

* * *

There are several loose ends flapping around in this chapter and I need to deal with the ones that have caught my eye. One loose end is whether there will be an end to events. When thermodynamic understanding of natural change emerged into the consciousness of its creators, they lit upon the prospect of the end of the world and a vision of its 'heat death'. That is not just climate change. Heat death, of the entire universe, is when disorder has burgeoned to the point that it is complete. In its last heat throes, all the energy of the universe, it is envisaged, would have degraded into chaotic thermal motion (colloquially, 'heat') and the opportunity for further increase in disorder, and therefore the opportunity for further natural change, will have been lost. All our structures, processes, achievements, and aspirations will be as though they had never been and there would be no prospect of the second law having a second chance.

Such may indeed prove to be our long-term featureless future. It is probably farther off than its predictors feared, for we know that the universe is not just a finite sphere but an enlarging and, we believe, accelerating balloon. Every day, as the universe expands, there is more room for chaos. It was once thought that the universe might one day collapse back to its pregnant initial pointlike egg, which causes worries about the future trajectory of entropy, but that prospect is no longer envisioned (nevertheless, we might be wrong on the timescale of quadrillions of years). I have already pointed out that our investigation of the universe has spanned only a few billion years, and a mightily different vision might be appropriate on timescales much greater than that (I have mentioned the possible circularity of time). No one yet has a clue about these matters.

Then there is the other end of things, their beginning. Can anything be said about the entropy, the measure of disorder, at the start of it all? If you accept my initial premise that not much happened at the Creation, then there is an answer.

Before the beginning (of this universe or perhaps some precursor original Ur-universe) there was absolutely nothing. This Nothing necessarily had perfect uniformity, for had it not it would not have been Nothing. If nothing much happened when Nothing rolled over into something, that perfect uniformity would have been maintained (I have supposed), and my thesis all along is that the new born universe would have inherited Nothing's uniformity. With no chaotic disorder, the initial entropy would have been zero.

The rest is history, literally. As time progressed, natural events that we would recognize as crumbling into disorder, globally but not necessarily locally, would have taken place. Stars would have formed, galaxies too. Planets came and went, biospheres and battles too. Thought, art, and understanding would have emerged certainly here, let's hope elsewhere as well, for it is too precious to be left only to us. We are still in the midst of this unwinding of the Creation, in the midst of increasing disorder, with local abatements that we call the attributes and artefacts of civilization.

Time's arrow thuds into this discussion, of course. The unstemmable rising tide of entropy lies in cahoots with the seeming irreversibility of time, providing us with a future and preventing us from revisiting and tinkering with the past. All our yesterdays are eras of lower global entropy and are unrevisitable (in many cases, thankfully). Given that there is time, and given that events occur that ineluctably are accompanied by an increase in entropy, only the future lies in store for us; the past is untouchable history frozen forever in its immutability. Yes, rising entropy, and in particular the local accumulation of experienced events, add to our memories and experience of the passage of time, but in that there should be nothing particularly mysterious given the arguments I have made about the process of sinking ceaselessly, unstoppably into chaos.

* * *

Or maybe not. In the background of all I have said lies a mystery relating to the laws of nature that it would be improper for me to conceal. All the fundamental laws of nature appear to be symmetrical in the direction of time. That is, deep down, Nature seems ignorant of time's direction, yet on the surface, in our experience, she is fully aware of its direction in the sense that the laws of nature either don't mention time or don't have a direction built in. That is, either the results are timeless (like the law of conservation of energy) or, when solved, run equally in either direction of time. An example is the solution of Newton's equations for the motion of a planet: you can trace its orbit into the future or, by changing the sign of time, into the past. There is nothing in the equations that insist you travel forward in time. How does our commitment to travel into the future stem from Nature's apparent indifference to the direction of time?

Boltzmann, the suicidal hero of Chapter 4, is relevant here but in a quite different context, although his contribution is not nearly as clear-cut as he had supposed. Boltzmann thought he had proved that if you start with an arbitrary collection of molecules whizzing around in accord with time-symmetrical laws, then regardless of the initial positions and velocities, they will settle down into the most random distribution. That is, he thought he had shown that time's asymmetry emerged from time's symmetry simply by considering the statistical outcome of the laws rather than focusing on the solution for a single molecule. He ascribed time's arrow to the behaviour of the crowd, not the individual.

To get the sense of his argument, think about one ball in one half of a box. It is moving, bouncing off the walls. There is a very good chance that its motion will bring it back to its starting point, at least fleetingly. Now consider two such balls. The complication now is that they might collide and bounce away from each other.

It is plausible, though, that you could still expect them to be back fleetingly in their initial arrangement, although you might have to wait quite a long time and it would also depend on the precision with which you assessed 'the same arrangement'. It is also plausible that you might find three balls back in the same arrangement, even four, but you would have to wait ages for the initial state to be revisited. But suppose you had 100 balls, even 1000, or trillions? In principle it might turn out that they revisited their initial state, but even with a mere 100 balls you might have to wait for the lifetime of the universe. There is thus a practical irreversibility even though the irreversibility is not present in the laws that determine the trajectory of each particle.

A more fundamental solution might lie more deeply in the warp and weft of the fabric of reality. What we take to be time's single arrow might in fact be the outcome of a coupling of two arrows of time, one statistical (the Boltzmann part), the other cosmological. The cosmological arrow of time might change even the 'in principle' reversibility. While you are hanging around with half an eye for nearly eternity on the 100 balls the universe has changed: it has got bigger. There is no such thing as revisiting the initial state, for with the swelling universe the spacetime that you started with and used to define the initial state of the system has become history and even in principle you cannot expect to find the balls in their original state, and the longer you wait the less probable it becomes.

Thus, although the laws of nature might be time-reversible, their manifestation in the real world of complex interaction and their staging on a changing cosmic arena have rendered them in practice time-irreversible. There is no going back.

* * *

I have dealt with three laws of thermodynamics, the zeroth (about temperature), the first (about energy), and the second (about entropy). There is a fourth law of thermodynamics, which inevitably because of that late-arriving 0 is known as the 'third law of thermodynamics'. Some people wonder whether it is really a law, because unlike the other three it doesn't introduce a new physical property. Perhaps that signifies that it merely rounds out the other three, completing thermodynamics once and for all.

The third law in the form originally proposed by the German chemist Walther Nernst (1864–1941) in 1905, amid a slight hint of argument about priority, effectively states that *absolute zero can't be reached in a finite number of steps.* If you were feeling sardonically gloomy, you could interpret the first law as asserting that nothing happens, the second law as asserting that if perhaps it does, then things get worse, and this third law as implying failure anyway. The Nernst statement resembles those of the second law by Kelvin and Clausius in the sense that it refers to observations, not an underlying molecular explanation. Deeper insight was obtained in 1923 when the two American chemists Gilbert Lewis (1875–1946) and Merle Randall (1888–1950) found a way to express the law in molecular terms, and asserted, in effect, that *all perfectly crystalline substances have the same entropy at absolute zero.* I cannot in these pages demonstrate why these two statements, so different in formalism, are identical in practice, but broadly speaking it arises from the fact that because all entropies tend to the same value, it takes more and more work to extract energy as heat as the temperature approaches zero, and an infinite amount of work finally to get there.[6]

All the third law states is that all substances have the same entropy at absolute zero. It doesn't reveal what that value is. However, Boltzmann's interpretation of entropy as a measure of disorder suggests a value: zero. Because the substance is a perfect crystal, all its molecules or ions are in perfect, serried arrays, so there is no disorder due to imperfections or a molecule lying in the wrong place. Because the temperature is zero, all its molecules are in the lowest possible energy level, so there is no disorder due to one molecule vibrating more than another. We are in the presence of perfection, which implies that the entropy is zero whatever the identity of the material. No wonder it is inaccessible!

The third law obviously has implications for those striving to reach very low temperatures and the fascinating physics they are hoping to find there. Even for the normal mortals who inhabit warm laboratories the law is essential, for the fact that 'entropy is zero' under certain conditions gives a starting point for a wide variety of calculations in thermodynamics, including the numerical prediction of whether a chemical reaction 'will go' or not. Such calculations are not likely to be of much interest in the context of this book, but you should be aware that the third law rounds out the other three and renders them more useful quantitatively than they alone would be.

I say rounds out the others; but could there be a fifth, sixth... law of thermodynamics? No one knows, although some would claim there are more to be found. Conventional thermodynamics, particularly the second law, deals with *tendencies* to change and systems that are at equilibrium, with no tendency to undergo still further change. There is some interest in formulating versions of thermodynamics that deal with the rate at which that initial tendency is realized, such as the rate of production of entropy in a process that is far from equilibrium and poised away from it, such as a living human body, where equilibrium is death. Such 'dynamical structures' were studied by the Russian-born Belgian chemist Ilya

Prigogine (1917–2003), and earned him the Nobel Prize in 1977; but certain aspects of his work, and his vision of its implication that determinism in nature is dead, remain controversial and to some, anathema, truly beyond anarchy.[7]

* * *

I have sought to show that Nature, left to itself, gets gradually worse but in the process throws up local abatements of chaos that might be exquisite. The second law of thermodynamics encapsulates this tendency of matter and energy to disperse, and gives deep insight into the undirected driving force that lies behind all natural phenomena. I think it extraordinary that such a simple everyday principle can account for all change. I have shown that its implications include the efficiencies of engines, and through that of economies, yet lurking beneath it is the problem of why Nature's time-symmetrical laws result in the unique flight of time's arrow. The second law is the child of anarchy, yet engenders the humbling and the amazing.

6

The Creative Power of Ignorance

How matter responds to change

Ignorance is an effective ally of indolence and anarchy. In this chapter I want to show how not knowing can be used constructively to arrive at knowing. The particular law of nature I want to elucidate initially was one of the earliest to be expressed quantitatively as scientists began to realize the importance of attaching numbers to Nature, but was understood only in the late nineteenth century. Its elucidation springs from ignorance.

The law in question concerns the structurally simplest form of matter, a gas, and was established by Robert Boyle (1627–91), working in Oxford in the early 1660s, or, as the French would claim, by Edme Mariotte (1620–84), working in Paris in 1679. It was elaborated by Jacques Charles (1746–1823) when, as so often happens, studies of Nature were stimulated by the demands and opportunities of technological advance, in his case by the rising interest in flying in balloons. Their laws are of historical interest because they are among the earliest summaries of the properties of matter expressed quantitatively, that is, in a manner open to numerical calculation and prediction. They are also the basis of the development of thermodynamics and its application to chemical and engineering phenomena and processes, and so are of great fundamental and practical importance.

I have already sketched the formulation of Boyle's law in Chapter 1, and need only mention its content briefly again here. Boyle, and independently Mariotte, for news was slow to travel in those days, established that *the pressure exerted by a gas is inversely proportional to the volume it occupies*. Thus, decrease the volume occupied by a gas and its pressure increases. Specifically, confine a gas to half its initial volume by pushing in a piston, and its pressure doubles. These days, we have little trouble in explaining the *qualitative* aspect of the law. The modern picture of a gas is of a swarm of molecules ceaselessly hurtling chaotically through empty space. When the gas is compressed, the same number of molecules are packed into a smaller volume but are hurtling around at the same average speed (because the temperature is constant, and temperature determines speed). As a result of the greater density of molecules, in any given interval more of them can strike the walls of the container. The force they exert on the walls is experienced as pressure, and because the total force is now greater, so is the pressure. The challenge, though, is to account for the *quantitative* aspect of the law, the numerically precise relation between pressure and volume when the temperature is held constant.

Charles added to this observation by examining what happened when the temperature is allowed to change. Flying in those days, the late eighteenth century, was initially achieved in hot-air balloons. A sheep, a duck, and a rooster were freight, and no doubt frightened, on their first flight on 19 September 1783 in a hot-air balloon built by the Montgolfier brothers, Josef and Etienne. Man took his first small, risky, but portentous step up into the air a few weeks later. Hydrogen balloons (soon after, the more readily available but less lift-producing and actually poisonous, inflammable, town-gas balloons) very quickly took to the air, being in those days more practical than hot-air balloons until modern-day bottled-gas-fuelled balloons reclaimed the skies in the 1950s. Gas-filled balloons avoided the need to have a flaming brazier as a companion in the basket of a hot-air balloon. Moreover, hot-air balloons could stay aloft only for as long as their heavy fuel lasted. In both types of balloon there was a serious interest as to how temperature affected air and thereby the lift capability of a balloon. For hot-air balloons, the lift came from the lowered density of hot air; the buoyancy of gas balloons depended on the temperature of the surrounding air, which varies with altitude. Another early investigator of the properties of gases, Joseph Gay-Lussac (1778–1850), and a colleague used a balloon in 1804 to set what was then a world altitude record of what they claimed with suspicious precision was 7016 metres (20 018 feet) in an intrepid attempt to analyse the variation of the composition and properties of the atmosphere with altitude.

In a series of experiments Charles, himself a pioneering balloonist, established what we now call *Charles's law*, that, when maintained at constant volume, *the pressure exerted by a fixed amount of gas increases in proportion to the temperature*. That is, double the temperature and the pressure doubles. You have to be careful here, because the 'temperature' of the law is the absolute temperature, the one normally reported on the Kelvin scale that I introduced in Chapter 4. It does not work for the more artificial Celsius and Fahrenheit scales. So if the gas is initially at 20 °C, think 293 K, and double that to 586 K (corresponding to 313 °C) to double the pressure the gas exerts; don't double 20 and expect the pressure to have doubled at a mild 40 °C.

Boyle's and Charles's laws can be combined into a single law, the *perfect gas law*, which in words reads *the pressure of a gas is inversely proportional to the volume and directly proportional to the absolute temperature.*[1] You will also come across the same expression called the *ideal gas law*: treat them as synonymous.[2] The law is 'universal' in that it applies to every gas regardless of its chemical identity and to mixtures, like air. Moreover, the mathematical form of the law depends on just one fundamental constant, unimaginatively known as the 'gas constant', and that constant is the same for every gas. The gas constant was in fact present but hidden in Chapter 4, for it is actually the more fundamental Boltzmann's constant in disguise. That disguise enables the gas constant to sneak into many expressions that have nothing to do with gases, for instance the expression for calculating the voltage of electric batteries.

In Chapter 1 I introduced the concept of a limiting law, one that becomes progressively more reliable as the substance being described is removed and applies exactly when all the substance has gone. The perfect gas law is such a limiting law, in the sense that it becomes more reliable as the pressure is reduced to zero or, equivalently, as the volume occupied by the gas approaches infinity. The perfect gas law describes 'perfect gassiness', the behaviour that would be observed if there were no complicating issues, such as the molecules briefly sticking together in the gas instead of flying around completely freely, or finding that the space available to one molecule during its chaotic hurtling through the container was already taken up by another molecule. When the volume occupied by the gas is very great, the molecules meet so infrequently (in the limit of infinitely vast amounts of space between them, never), that they are oblivious to each other's presence. In practice, that means that the perfect gas law is obeyed perfectly in the limit of infinite volume or zero pressure. In short, the perfect gas law is obeyed exactly by all gases only when they are not there.

Despite that last remark, limiting laws are far from useless. Like many limiting laws, it turns out that the perfect gas law is a sensible starting point for a discussion of actual systems, in its case, of all gases under everyday conditions, with individual idiosyncrasies showing up and needing to be accommodated only as more complex behaviour is identified. That's a bit like the 'perfect' route between two places being a straight line, as crows are said to fly, which may be a sensible way to consider a journey initially, but in practice you have to stick to the actual nearby roads. The straight line is the 'limiting' route in the absence of the various interfering features of the landscape. In practice, it is found that the perfect gas law is reliable at the normal pressures encountered in real-life applications, but deviations have to be incorporated for very precise work, when pressures are abnormally high, or when temperatures are so low that a gas is not far off condensing to a liquid. In fact, the perfect gas law is of huge importance because it underlies the formalism, in the sense of being a starting point for the formulation of expressions, of much of thermodynamics and its applications.

There are several limiting laws in science, all of them expressing the 'perfect essence' of a particular property, and all of them act as a sensible and useful starting point for describing more elaborate behaviour. The ones I have in mind, but will do no more than mention, concern the properties of mixtures of liquids and the effect of dissolved substances on the properties of solvents. Most of them were identified in the early days of serious chemistry and bear the names of their discoverers, who include the English chemist William Henry (1774–1836) concerning the solubility of gases in liquids, such as the manufacture of soda water and champagne and the incidence of 'the bends' in deep-sea divers, the French chemist François-Marie Raoult (1830–1901) concerning how a dissolved substance affects the properties of the solution, such as the effect of salt on the freezing point of water, and the Dutch chemist Jacobus van 't Hoff (1852–1911) concerning the life-affecting property of osmosis (a word derived from the Greek for 'push'), the apparent ability of a solvent to push its way through a membrane and which, among other things, keeps biological cells fat and healthy, plants unwilting, and trees fed with nutrients.[3] Apart from illustrating how easy a sufficiently early academic bird can achieve the worm of immortality through the identification of simple systematic behaviour, and perhaps the agreeably international character of the scientific enterprise, none of these laws is of great relevance to the current discussion despite being of considerable relevance to physical chemists like me. All three laws can be traced back to the anarchic collapse into chaos captured and expressed by the second law of thermodynamics. In a nutshell, plants would wilt, crops would fail, and you and I would die in the absence of these simple consequences of the second law. They would be truly limiting.

* * *

With those apologetics out of the way, it is time to see how ignorance augments indolence and anarchy to illuminate the perfect gas law, the initial and principal target of this chapter. You already know that a gas consists of molecules in a state of ceaseless chaotic motion, zooming hither and thither, striking each other and moving off in a new direction and different speed. You and I stand in a storm of molecules even on the calmest of days, our surface being battered by this ceaseless tempest and more or less holding us in shape. To imagine the scale of the turmoil in a gas like air, think of a molecule as being the size of a tennis ball zooming through about the length of a tennis court before striking another molecule. All this chaotic activity is taking place under the regime of classical mechanics, the outcrop of indolence and anarchy.

The ignorance from which knowledge will spring is the total lack of knowledge we have about what is happening to an *individual* molecule in this maelstrom of ceaseless activity among myriad others. Although each individual collision contributes to the pressure, there is no need to follow the trajectory of every individual molecule. Just as in sociology it is possible to predict with reasonable confidence the behaviour of a crowd but not an individual within that crowd, because myriad molecules are present in the crowd we call a gas, we can be confident about disregarding the contribution of individual molecules simply because we are ignorant of them. Like quantitative sociologists, we need to stand back and focus not on the individual but the crowd.

When that is done, and the mathematics of the swarm is carried through by using Newton's various laws, out pops Boyle's law.[4] You might think this is so obvious it doesn't need a lot of mathematical paraphernalia to reach that conclusion. In a sense that is correct, but in fact the mathematical outcome is richer than just Boyle's law, for it shows how pressure and volume jointly depend on various features of the gas, such as the mass of one of its molecules and the average speed of the molecules, a dependence which could, I think, not be inferred from the pictorial interpretation of the law. Once again, the application of mathematics to elucidate a law by expressing a qualitative picture quantitatively enriches insight and understanding.

* * *

What about Charles's law, that the pressure is proportional to the absolute temperature? Once again, ignorance in alliance with (under-the-counter) mathematics will help, and will result in an explanation of his law.

The key is the connection between molecular speed and temperature. Speed plays a dual role. More molecules can bang into the wall in a given interval if they are moving fast, and when they bang they deliver a greater impact. Therefore, as the temperature is raised the pressure exerted by the gas should increase on both accounts, both due to the frequency of impacts and the strength of the blows they deliver. It is known (I'll return to the point in a moment) that the average speed of molecules in a gas is proportional to the square root of the temperature (the absolute temperature), so because two square roots of the temperature when multiplied together give the temperature itself, we come to an explanation of Charles's law, that the pressure is proportional to the temperature.[5]

The outstanding question is why the average speed of molecules in a gas is proportional to the square root of the temperature. First, let's see what that means in practice for air. The average speed depends on the mass of the molecules, with light molecules (nitrogen) hurtling along at 500 metres per second (1120 miles per hour) at 20 °C and heavier molecules, such as carbon dioxide, lumbering along at 380 metres per second (850 miles per hour). Those figures actually give some insight into another physical phenomenon: the propagation of sound. The speed of sound in air is about 340 metres per second (760 miles per hour) at sea level, which is similar to these molecular speeds. Sound is in fact a pressure wave that depends on the molecules of air adjusting their positions collectively to give a wave of undulating pressure, and the rate at which they can do so depends on the speed with which they can move. Not surprisingly, therefore, the speed of sound is comparable to the average speed of molecules. Our focus here, though, is on the dependence of molecular speed on temperature, and because it varies as the square root of the temperature, it's easy to calculate that the average speed falls by about 4 per cent on going from a mild day at 20 °C (298 K) to a cold day at 0 °C (273 K).

I need to account for the fact that the average speed of molecules is proportional to the square root of the temperature, for then we shall have accounted for Charles's law fully. Gas molecules have only kinetic energy, the energy due to motion, because in the limit of being far apart most of the time they do not interact with one another and so have no potential energy arising from their relative locations. Speed is related to kinetic energy, the energy of motion, so a backdoor into the calculation of average speed is to evaluate the average kinetic energy of the molecules and then to interpret that average energy in terms of average speed. In Chapter 4 you saw that Boltzmann has shown how to calculate average properties, by imagining tossing books (molecules) on to shelves (energy levels) and identifying the most probable outcome without guiding it in any way (except by ensuring that the total energy had a fixed value). When that procedure is carried out for a gas, an expression for the average kinetic energy of the molecules, and hence from that their average speed, drops out of the calculation. Sure enough, that average speed is proportional to the square root of the temperature, just as we need to account for Charles's law.

That's not all, of course. You have already seen that a combination of Boyle's and Charles's laws is the perfect gas law, the limiting law that is a starting point for so many of the applications of thermodynamics. You now know its origin: the Boyle's law component comes from a consideration of the number of impacts from the molecules of the gas, and the Charles's law component comes from the role of molecular speed and its dependence on temperature.

I hope you can appreciate the breathtaking nature of this conclusion. From ignorance, in this case not knowing any details of individual behaviour, has been distilled a law of nature, the perfect gas law. Along the way has emerged an understanding of the average speed of molecules in a gas and its variation with temperature and (although I have not mentioned this feature except in the Notes) on the mass of the molecules making up the gas. Ignorance appropriately marshalled and deployed can be a powerful source of understanding.

* * *

The perfect gas law is just one of the little laws of nature that I mentioned in Chapter 1. They are dependent laws, these outlaws, hanging as they do like fruit from the trees of the great laws, the inlaws. Maybe there are other laws that once the mother laws have been accepted as consequences of indolence and anarchy, emerge from a further application of ignorance.

One such outlaw is Hooke's law, which I also mentioned in Chapter 1. Robert Hooke (1635–1703) was one of the truly imaginative thinkers of the seventeenth century, as bewilderment and obfuscation were driven back by the incoming tide of rational thought represented by the Enlightenment, feeding as he did into Newton's thinking and generating great thoughts of his own. As we saw in Chapter 1, Hooke's law says that as you stretch a spring, *the restoring force is proportional to the displacement*. So, pull a spring out a centimetre from rest and feel it resisting, pull it out twice as far and feeling it resisting twice as powerfully.[6] A consequence of this law, a consequence that stems directly from Newton's anarchy-based mechanics, is that springs oscillate steadily as do pendulums, and consequently the clockwork world keeps time.

Hooke's law is another example of a limiting law, being strictly valid only if there is no displacement from equilibrium. It is exactly true of stretched springs, provided they aren't stretched, and exactly true of swinging pendulums, provided they aren't swinging. All springs and pendulums show deviations from the law for measurable stretches and swings, but they increasingly conform to the law as the stretches and swings approach zero. In most cases the deviation is negligible and the law can be used to make reliable predictions and clocks keep excellent time. But stretch too far and, like overstretched elastic, the law snaps.

Where does ignorance step in to account for Hooke's law? If you were ignorant of external factors, then you would come to Hooke's conclusion wherever there is a force that opposes displacement. The argument runs as follows. Think of any property, and then think of how it changes as a parameter is modified. By 'parameter' I mean anything you can tweak, such as the extension of a spring, the angle of a pendulum, the length of a bond in a molecule, the pressure applied to a lump of solid, and so on. In each case there might be a property of the object being tweaked that depends on the magnitude of the tweak and reaches or passes through a minimum at zero tweak. For instance, it might be the energy of a spring that is being extended or compressed. If you were to imagine a graph being plotted of the energy of the spring against the tweak, it would look like a curve that rises up on each side of zero tweak. Thus, the energy of a spring rises as it is either extended or compressed, and is a minimum when it is at rest.

Unless there is very odd behaviour coming into play, all such curves, whatever the property and whatever the nature of the tweak, start off in the same way from the value at zero tweak. That manner of starting out for the dependence of the energy on the displacement, is shaped like a parabola, and implies (according to Newton's mechanics) that the restoring force is proportional to the displacement, exactly as Hooke's law asserts.[7] Thus, not knowing how the force behaves results in us knowing how force is most likely to behave.

* * *

I would like to return to the point I made a moment ago about how Hooke's law underlies the world's timekeeping, or at least did when clocks were governed by swinging pendulums and watches were governed by oscillating balance wheels. Is it possible to identify a link between a restoring force that is proportional to a displacement and the regular beat of oscillation needed for timekeeping without solving the equations of classical mechanics? Is there a deep, perhaps unsuspected, explanation underlying regular oscillation?

Regularity in either space or time suggests an underlying symmetry. We need to identify it. In this case, on account of the dependence of the kinetic energy on the speed (and therefore the linear momentum) of the swinging pendulum's bob and of the potential energy as the bob rises and falls under the influence of gravity, there is a regular flow from one form of energy to the other and back again. One, the kinetic energy, depends on the square of the linear momentum; the other, the potential energy, depends on the square of the displacement.[8] Each form of energy feeds off the other, and the outcome is a regular exchange. As we watch the swing of a pendulum we see it pause at its turning point at the limit of its swing where it has zero kinetic energy but maximum potential energy. That potential energy feeds back into its kinetic energy as it accelerates. When briefly vertical, the pendulum is moving fastest: it has lost all its potential energy and now has maximum kinetic energy. Now that kinetic energy feeds back into potential energy as the pendulum slows but rises on the other side of its swing. The symmetry of the flow persists regardless of the width of the swing, and the pendulum beats as the heart of a clock.

Here is the point I want to make in order to burrow more deeply into this symmetry, but it needs a little setting up. There are two ways of looking at the world. It can be described in terms of the locations of things, or it can be described in terms of the momenta of things (I introduced linear momentum in Chapter 2, as mass times velocity). My description of Nothing has so far been in terms of position: I said that Nothing is obviously uniform, for localized bumps and shallows in absolutely nothing is inconceivable, or at least an oxymoron. But suppose you put on a different pair of glasses and look at Nothing in terms of momenta. This process is by no means as outlandish as it might seem, for there are techniques of investigation of matter that do just that. For example, many of the marvellous insights about biology (the structure of DNA, for instance) have been obtained by doing so. I have in mind the technique of 'X-ray diffraction', in which a beam of X-rays is passed into a crystal and emerges scattered (technically, diffracted) to give a pattern of spots that analysis turns into information about spatial arrangement. The spots are essentially the structure of the molecule looked at through momentum-spectacles.[9]

You can probably accept that when you look at Nothing through your new glasses you still see nothing. This new vision of Nothing is just as uniform as the old position-based vision revealed. That being so, and if you continue to accept that not much happened at the Creation, then immediately after the incipience of the universe it remains uniform in terms of linear momenta. A consequence of that uniformity is that the laws of nature are independent of the velocities ascribed to everything. Provided the observer and the observed are travelling at the same velocity (so we don't have to worry about relativity), the laws are the same. For instance, the swing of a pendulum obeys the same law whether the clock it drives is travelling at 100 metres per second or is stationary. (You would report deviations if it was stationary and you were moving, but that is relativity and an entirely different kettle of fish: see Chapter 9.)

The universe has a deep symmetry when described in terms of positions and linear momenta. A pendulum displays this symmetry. The energy of a swinging pendulum (indeed, any oscillator, including a weight on the end of a spring) has equivalent contributions to its energy that are symmetrical in space and momentum. As it swings, energy flows from its linear momentum into its displacement, then back again. If you took off your everyday glasses and put on your momentum-spectacles, you would see no difference. Where once you saw displacement, now you would see momentum, and vice versa. It is this symmetry that underlies the rhythmic, ceaseless swing of the pendulum, the weight on the end of a spring, or the ceaseless oscillation of the balance wheel.

* * *

Ignorance has taken us a long way in this chapter and has emerged, suitably channelled, as the basis of knowledge. One kind of ignorance in question is the absence of knowledge about the behaviour of individual entities, forcing us, like sociologists, to resort to assessing the behaviour of the crowd, in our case a gas of molecules, and finding that the gas obeys certain laws that are then put to use with great effect in thermodynamics. Then there was a second kind of ignorance, about the behaviour of entities—I have in mind pendulums and springs, but much else too—when displaced a little from rest, which, except in special cases, are likely to behave in the same characteristic way, and are manifest as laws that account for their behaviour. That behaviour, as a bonus, also turns out to exhibit another deep symmetry of the universe that springs from its inception from uniform Nothing.

7

The Charge of the Light Brigade

The laws of electricity and magnetism

The laws of electricity and magnetism play a special role in Nature, not merely for the ways in which they underlie our existence and structure but also because they are the foundations of most of the businesses, pleasures, and pastimes of life. The radiant energy of the Sun is brought to us as electromagnetic radiation and where it lands it drives photosynthesis to build our biosphere, coating the inorganic surface of the Earth with forest, field, and prairie and greening the sea. Through that invigoration it populates land, sea, and air with mobile organisms and ultimately provides our breakfasts, and through our breakfasts our creativity and enjoyment. More fundamentally, electromagnetic forces hold atoms and molecules together and are responsible for the existence of multifarious forms of tangible matter. Those forces are also responsible for transport and communication, which range from the essential to the trivial but have become an ineluctable part of human enjoyment, slaughter, commerce, persistence, and, in a word, existence.

I have slipped in another aspect of the importance of these laws by segueing from electricity and magnetism into 'electromagnetism', their unification. The unification of all the forces, showing that they are all a manifestation of a single mother force, is one holy grail of physics. Its achievement, currently far from fully complete but seductively beckoning, would show that a seemingly motley collection of forces has an underlying unity and that the world is simpler than appearances, taken at face value, currently suggest. The corresponding motley collection of laws that describe the operation of these various forces would then blend into a single law, and its origin would be more likely to be discoverable.

Symmetry plays a central role in the search for the grail. I have touched on its role in other contexts in earlier chapters. There was Emmy Noether's identification of the deep connection between symmetry and conservation laws in Chapter 2, my view that the uniformity of Nothing survives the birth of the universe, and most recently my vision of clocks as the public face of a hidden symmetry permeating the underworld of properties.

If you would like a concrete analogy of the role of symmetry in the unification of electricity and magnetism (and by extension, of the other forces), then the following might help. Think of a square as representing electricity and a regular hexagon as representing magnetism. These two shapes are quite distinct and there is no way of twisting and turning one to make it into the other. Square electricity and hexagonal magnetism seem to be distinct. Now think of a cube. When you look at a cube face on, you see a square. When you look at a cube along a body diagonal (which runs through opposite corners), you see a hexagon. Now, if instead of thinking of the two forces as separate entities, we should enlarge our vision, step up a dimension into three, and think of them combined into a cube, then square electricity and hexagonal magnetism are clearly the manifestation of a single entity and are related by rotation in some kind of abstract space. The cube is electromagnetism.

I have more to say in the same vein and as this chapter unfolds it might be helpful for you to think of an extension of the cube analogy. As I have mentioned, a major thrust of modern theoretical physics is to unify all the forces of the world and in particular to demonstrate that the unified electromagnetic force is but one face of a single force. This further unification has already been achieved with the 'weak force', the force that operates within atomic nuclei, twisting their components apart and resulting in radioactivity, the radiation spat out from a nucleus either as gamma rays (very short-wavelength, highly energetic photons of electromagnetic radiation) or charged particles (alpha and beta rays). Attention is currently aimed at certain aspects of the 'strong force', the very short-range force that binds the components of a nucleus together despite the electromagnetic force striving to drive the closely packed electrically charged particles asunder. We should be grateful that the strong force does not have as long a range as electromagnetism, for if it had, we and everything there is would be sucked down into a single colossal atom. Maybe gravitation will join in the unification one day, although it does seem to have some enigmatic features related to the structure of spacetime itself. Unification of all the forces will have been achieved when the elaboration of the cube and its rotations and other manipulations in some kind of elaborate abstract space have been identified. Unification is the hunt for a grail; that grail is not a cup: it is a currently unimaginably elaborate cube in an abstract space of many dimensions.

I shall concentrate on the laws of electromagnetism in this chapter, largely because they will be the most familiar, or at least the least unfamiliar. (I shall review bits of them, anyway, to distil their essence.) Similar remarks will apply to the weak (certainly) and strong (I think) forces, so my aim is to identify the origin of these familiar laws and leave it to your imagination to accept that similar remarks apply to the other forces. That, I must admit, is a leap of faith, for physicists are still struggling to incorporate them into a unified scheme.

In case you lose sight of where I am leading, let me remind you that I want to argue that the laws of electromagnetism are yet another consequence of indolence and anarchy, stemming from just about nothing happening at the incipience of the universe, when something emerged essentially and presumably spontaneously from Nothing. There are several aspects that we need to think about.

* * *

One of the earliest laws of electromagnetism to be formulated was *Coulomb's law*. The French physicist Charles-Augustin de Coulomb (1736–1806) formulated the law that bears his name in 1784, and proposed that the force between two electric charges weakens as the square of the distance between them. This is a so-called 'inverse square law'. The same conclusion had been arrived at by others (including Joseph Priestley and Henry Cavendish, both in England), but Coulomb is generally credited with a systematic investigation and formulation of the law. There had been earlier hints that the inverse square dependence on separation was likely, for gravitation obeys the same dependence.[1]

My suspicion is that, if you were God and, despite your apparently capricious decisions in other instances, wanted to give humanity the most beautiful law of interaction of electric charges that you could devise and they could appreciate, you would give them

Coulomb's law. It is a law of singular beauty, but that beauty is more than skin-deep and unseen by the eye of the casual beholder.

First, and rather trivially, the force it describes is spherically symmetrical, like a ball, that most perfectly symmetrical of bodies in three dimensions. I shall explain that point in a moment. Can the 'most' in that assessment be quantified? Yes: a sphere has an infinite number of axes of symmetry (any diameter), and an infinite number of angles of rotation about any of those axes leaves the ball apparently unchanged. Now think of a mirror slicing through the centre of the sphere and reflecting one hemisphere into the other: that mirror can lie at an infinite number of different orientations. In three dimensions, there is no object that possesses a higher symmetry: a sphere is the most symmetrical three-dimensional object, with its plenitude of infinities. If you are inclined to identify symmetry with beauty, then a sphere is the embodiment of beauty, or at least of some kind of perception of primitive beauty.

Coulomb's law is spherical in the following sense. The direction of one charge relative to another is irrelevant. The strength of the interaction falls off with distance in the same manner and to the same extent in any direction. That might not seem particularly exciting, but it has profound consequences for the structures of atoms and the consequent properties of matter. Furthermore, here lies another hint about the importance of Nothing. A simplistic view (but one that might be correct, nevertheless) is that the spherical symmetry of the Coulomb force emerges from the uniformity, and specifically the spherical symmetry, of absolutely nothing. When the force emerged (and I shall discuss that in a more sophisticated way later in the chapter), Nothing evolved into a medium for the propagation of the force, and its emergence from absolutely nothing did not have imposed upon it any additional constraint. Indolence lies in Coulomb's heart.

Second, despite what I have said about a sphere being of infinite symmetry, Coulomb's law is more than spherically symmetrical. It has an inner symmetry not discernible by simply seeing distance in its expression and no mention of direction. If we let the eye of the casual beholder become more sophisticated, and instead of assessing the interaction in our common or garden three dimensions, we step into a fourth, then that spherical symmetry is preserved there too and becomes what I shall call hyperspherical.[2]

I realize that I am probably asking your visual imagination to exceed its bounds (as I am of mine too), but such is the power of mathematics that it can take that step, and demonstrate symbolically that what I have said is true. I can give you a pictorial hint about what is involved in stepping into four dimensions by taking a step from two to three dimensions and then asking you to accept that something similar happens when you step up to four. You have already seen that a square and a hexagon can be shown to be related by moving from a plane in two dimensions to a cube in three, and I am asking you to think in a similar way here, although the issue is a bit different.

Here is the image I want you to have in mind. Think of a square sheet of paper with a big red circle at its centre. Now think of the same sheet of paper with one half coloured red and the other half untouched. Clearly, there is no relation between the two patterns. Or is there? A square and a hexagon became related when I invited you to step up a dimension and think of a cube; might this also be true when we do the same to the circle and the rectangles?

I need to invite you to step from two into three dimensions by imagining a sphere resting on the centre of the sheet of white paper. Let the sphere have its southern hemisphere coloured red and its northern hemisphere left white. Imagine drawing a line from the North Pole through the sphere and on to the paper. If the line passes through red, the paper below is coloured red. As you can probably visualize, that process results in a red circle centred on the point where the South Pole rests on the paper. Now rotate the sphere through 90° around an axis passing through its equator, so now its western hemisphere is the red half. Repeat the projection exercise from the new North Pole at the top of the sphere. As you can now probably visualize, one half of the paper now becomes red and the other remains white. You can now see that although in two dimensions the two patterns are unrelated, they are related if you step up to three: there is a single sphere, and our perception in our two dimensions has failed to identify the underlying symmetry. So it is with the Coulomb law: only if we step into four dimensions can we appreciate its full symmetry, the same in every direction in four dimensions.

There are not, as far as I am aware, any immediately interesting *everyday* consequences of this hidden, higher-dimensional symmetry. I am aware of one consequence, which is an esoteric aspect of the structure of hydrogen atoms, which I am reluctant to go into despite those atoms constituting (apart from dark matter) the most abundant matter in the universe.[3] I could stretch a point and say that the elimination of that hyperspherical symmetry of the interaction of an electron with a nucleus when more than one electron is present in an atom (as is the case for all elements other than hydrogen) is responsible for the structure of the Periodic Table of the elements and the entire world of chemistry (which includes biology and, by extrapolation, sociology). But that would be little more than showing off; nevertheless it should be kept in mind.

There is, in fact, in relation to this point an opportunity to introduce another rather vague law of nature, the *periodic law*, which recognizes that the properties of the elements repeat each other to a certain extent, after a variety of intervals when arranged according to their atomic number (the number of protons in a nucleus). Thus, silicon (atomic number 14) resembles carbon (atomic number 6) eight elements earlier, and chlorine (atomic number 17) resembles fluorine (atomic number 9) eight elements earlier too. The periodic law, the Periodic Table, and the totality of chemistry, biology, and sociology, is a direct consequence of the symmetry of the Coulomb interaction in alliance with rules about the way that negatively charged electrons congregate around the positively charged nuclei of atoms.

* * *

I shall now dig deeper into the Coulomb interaction and the other interactions that are responsible in some cases for holding matter together and in others for driving it apart, and look more deeply for their origin in indolence and anarchy.

My starting point is the Schrödinger equation for the propagation of waves, which I introduced in Chapter 3. I mentioned there that a central aspect of quantum mechanics is the 'duality' of matter, that particles take on aspects and waves, and vice versa. The point in this connection that I need to make is an interpretation of a wave that is due to the German physicist Max Born (1882–1970; his various contributions to the formulation of quantum mechanics were belatedly recognized with a Nobel Prize in 1954). The 'Born interpretation' of a wave in quantum mechanics is that the square of its amplitude in a region tells you the probability of finding the particle there. With that interpretation in mind, think of a wave of uniform amplitude (the height of its peaks) and constant wavelength (the separation of its peaks) stretching from here to the horizon and beyond. Now think of shifting the whole wave along a bit, so that all its peaks and troughs are moved a little. Nothing observable has changed, in the sense that if you were to evaluate the probability of finding the particle at any point, then you would find the same result.[4] We say that the observation is invariant (that is, not changing) under a global (that is, everywhere the same) 'gauge transformation'. That last term needs only a few words of explanation. All it means is that if you laid a measuring rod (a gauge) along the wave, and noted the location of the first peak of the wave, then after the wave has been nudged along a bit, you would need to move the gauge (the rod) along a little in order to get the same reading.

So far, everything is a bit trivial, perhaps like all ultimate truths, but we are on the threshold of entering the world of the 'gauge theories' of particle interactions, one of the frontiers of modern physics. Now for the detrivialization of what has gone before, which will prove quite startling in its consequences.

I explained, or at least mentioned, in Chapter 3 that the equations of motion of particles can be established first by setting up an expression for the 'action' associated with a path, and then looking for the path that involves least action. That path of least action is the path the particle takes, being the one that survives, unwashed away by its neighbours. I went on to say that Newton's 'differential' equations can be seen to be a way of instructing the particle on how to grope its way along that path, infinitesimal step after infinitesimal step. That discussion was expressed in terms of actual 'familiar' particles, such as electrons, but it also applies to the less tangible particles of electromagnetic radiation, photons, because in quantum mechanics everything is both a particle and a wave. Therefore, the principle that a particle adopts the path of least action can be applied to electromagnetism and its propagating particle, the photon.

In the case of electromagnetism, an expression for the action is formulated, it is minimized, and out of that minimization drops the equivalent of Newton's equations, but now the equations describe the behaviour of the electromagnetic field. These equations are known as the 'Maxwell equations', and were formulated by that briefly incandescent shooting star James Clerk Maxwell (1831–79) in 1861. They were a mathematical version of the pioneering experimental investigations made by Michael Faraday (1791–1867) at the Royal Institution in London into electricity and magnetism. The equations demonstrated the interrelation between 'square' electricity and 'hexagonal' magnetism and effected their unification as 'cubic' electromagnetism. A hint about how to visualize that unification is to realize, as I explain in more detail in Chapter 8, that according to the theory of special relativity the outcome of moving is to rotate what you think of as space into time and vice versa. The faster you move, the greater the rotation, and what began looking like the 'electric' square face of the cube increasingly looks like the 'magnetic' hexagonal shape, and vice versa.

The Maxwell equations are essentially a summary of the unified laws of electricity and magnetism, so once we know where the equations come from, we shall know where those laws come from too.

* * *

In the late 1700s the Italian mathematician Joseph-Louis Lagrange (1736–1813; he is always so known, but was born Guiseppe Lodovico Lagrangia and became frenchified on account of his long sojourn in

Paris) had formulated a particularly elegant version of Newton's mechanics, which remains ideally suited to the development of the equations of motion we are looking for. His procedure involves some constrained guesswork. First, subject to various technical considerations I needn't go into, you need to propose a mathematical function known appropriately as the 'lagrangian'. There are various rules for writing down a lagrangian, one of which is that when it is used to evaluate the action, and then the action is minimized for the path between two points, the resulting expression is the experimentally verified equations of motion, in this case the Faraday-inspired Maxwell equations. If the minimized action doesn't conform to the known laws of motion, your guess about the form of the lagrangian was wrong, and you need to go back to the beginning and start again, and to do so until you get the Maxwell equations.

It turns out that this string of steps—*lagrangian* → *action* → *minimize* → *Maxwell equations* → *Faraday's experiments*—is achieved if the lagrangian is expressed in terms of a wave and that wave has a special relation to the electromagnetic field it describes. Here is the main point. We are free to move that wave backwards and forwards along its line of flight (that is, change its gauge), but because there can be no physical effects stemming from this change, the lagrangian cannot change, for otherwise the equations of motion, the Maxwell equations, would no longer agree with observation. That is, the lagrangian must be globally gauge invariant.

Now it is time for Noether and Lagrange to marry. You will recall from Chapter 2 that Noether identified a relation between symmetry and conservation. Global gauge invariance is a symmetry, so there must be an associated conservation law. It turns out to be *the conservation of electric charge*. That is, electric charge can be neither created nor destroyed.

I can give you a hint about how this conservation springs from global gauge invariance. Think of a little transparent cube embedded in the region occupied by the wave. When the wave is nudged along a little (its global gauge is changed by the same amount everywhere) some of it flows into the cube through one face and some flows out through the opposite face. For the lagrangian in the region to be invariant (wherever the region is located, and overall too), any mismatch of the inward and outward flows must be compensated by the creation or annihilation of amplitude inside the cube. This is the standard interpretation of a 'continuity equation', which is the mathematical form of the statement that the net flow through the walls of a region must be equal to the rate of change of creation or destruction of charge within the region. That is, charge is conserved.[5]

I would like to take two more steps in this discussion. First, I think it arguable that when the universe tumbled into existence and not much happened, there was no preselection of phases of the waves (the relative location of their peaks) that in due course will turn out to be the basis of electromagnetism. That is, when humans stumbled into the equations that describe electromagnetism, there was no requirement for them to identify and adopt a particular gauge: any gauge should work. In other words, as a result of indolence at the start of the universe, the equations of electromagnetism are globally gauge invariant, and as a consequence electric charge is conserved.

The second point is as follows. If as a result of indolence at the Creation electric charge is conserved, and the universe is stuck with what it has got, and always has had and always will have, a natural pair of questions to ask are how much electric charge there is in the universe, and how did that charge emerge from absolutely nothing.

We can be confident about the answers to one of these questions. The net charge of the universe is zero: there is certainly a lot of positive charge (think of all the atomic nuclei there are) and a lot of negative charge (think of all the electrons there are), but they cancel: the total charge is zero. We know that the charges cancel because the strength of the interaction between electric charges is vastly stronger than the gravitational attraction between masses so that even a slight imbalance of charge would have resulted in the universe blasting apart as soon as it had formed, with gravitation impotently standing by, overwhelmed and unable to hold it all together. We have to conclude that although there are equal and opposite charges, there is the same amount of both. The corollary is that charge did not have to be created out of Nothing, a Nothing where there was no charge initially. All that had to happen was for Nothing to separate into its electrical opposites.

Now, I have no idea how that took place, but it seems to me that the task of accounting for the separation into opposites is conceptually (and maybe practically) much simpler than explaining the actual creation of charge. After all, it is easier to understand the presence of a heap and a hole than a heap alone. The separation of Nothing into electrical opposites will have to be explained one day, but that is probably easier to achieve than finding a mechanism for the actual creation of two types of charges.

* * *

Here is another really interesting, perhaps astonishing, point. Suppose instead of a global gauge transformation where the whole wave is shifted as one, we make a *local* gauge transformation, with the shift in the wave being different at every point on its length. So at this point the peak of the wave is advanced a bit, at another point it is advanced a bit more, at another point it is shifted back a bit. Have in your mind's eye a wave that is distorted, scrunched up, in a local rather than a uniform global way.

If the wave is still to describe the same physical reality, the Schrödinger equation has to look the same. But unless you are careful it doesn't, because the shift in the peaks introduces extra terms. However, the extra terms can be eliminated by modifying the Schrödinger equation by adding a term that represents a new contribution to the energy. What is happening is that the shift of the peaks affects the energy of the original state, but the new term subtracts that change in energy (and does a couple of other technical things which ensure that the wave represents an actual energy state).[6] The astonishing conclusion, which is drawn by examining the form of the new equation, is that the new term in it represents the effect of an electromagnetic field. That is, the requirement that the Schrödinger equation is locally gauge invariant (that is, the description of nature remains the same under local modifications of the wave) implies the existence of electromagnetism and its Maxwell-inspired equations. Electromagnetism and its laws have emerged from symmetry, in this case local gauge symmetry.

You can probably see where this is going in relation to the emergence of laws of nature at the Creation. Given that not much happened (as usual, I would like to say 'nothing at all', but that is beyond my grasp at the moment), the uniformity of absolutely nothing was preserved as the universe slipped into existence. Now, though, the uniformity of Nothing was preserved in a much more subtle way than merely by resulting in uniform spacetime. Now it was preserved in a way that corresponds to the inner workings of space and time: to local gauge invariance.

Indolence then results in electromagnetism. By electromagnetism I then mean all the discoveries of Michael Faraday in his cellar laboratory at the Royal Institution and all their mathematization by James Clerk Maxwell and the laws of optics that follow from his recognition that light is an electromagnetic wave. Rarely has utter indolence achieved so much, including as it does so many of the attributes of modern civilization as manifest in communication, computation, transport, industry, commerce, entertainment, and the general comforts of life.

* * *

Why stop at electromagnetism? After all, it is known that there are several types of force operating in nature. Electric charges interact by electric forces and moving electric forces interact by magnetic forces. Maxwell's work showed that these two forces are manifestations of a unified 'electromagnetic' force and we have seen that they emerge from a certain type of local gauge invariance. The so-called 'weak force', so called because it is so much weaker than the electromagnetic force under the present conditions of the universe, is responsible for certain kinds of nuclear decay, prising elementary particles apart and resulting in them being flung from nuclei as radiation.

Then there is the 'strong force' which binds the components of certain elementary particles so strongly together (hence its uninspiring name) that it overcomes the electromagnetic force that otherwise would drive them asunder. Thus it is that nuclei persist despite the fearsome power of the electromagnetic force that threatens to destroy them but which is overcome by the strong force until the feeble weak force ferrets away in the background and in due course might triumph. It is as a consequence of the strong force that a whole variety of different nuclei exist. Those nuclei, as a result of their positive electric charge and the attraction between opposite charges, capture negatively charged electrons to give rise to electrically neutral atoms. Those captured electrons are held to their mother nuclei only quite weakly and can be removed by quite gentle battering, and thus give rise to chemistry, and through chemistry biology, and through biology zoology, and through zoology sociology, and through sociology civilization. Finally there is gravitation, perhaps the subtlest force of all, that binds everything, but so weakly yet universally and results in galaxies, stars, solar systems, planets, and the late attainment and achievements of human life.

All these forces (and perhaps others yet unidentified) account for the panoply of existence, the fabric of reality, the complexity of existence. All of them can be expressed as arising from local gauge invariance and hence as the consequence of indolence. There are, admittedly, severe problems with this view.

One difficulty has been overcome. The local gauge invariance that gave rise to electromagnetism was of a very simple and readily visualizable kind: it took a wave and scrunched it up in ordinary three-dimensional space. For technical reasons, this kind of scrunching is regarded as an example of an 'Abelian' gauge transformation. Niels Abel (1802–29, struck down by tuberculosis) was a Norwegian mathematician who, among other important matters, studied symmetry transformations that ended up with the same result regardless of the sequence in which they were carried out (rotate left, reflect, then rotate right, for instance). The scrunching that has been needed to show that the weak and strong forces arise from a variety of local gauge invariance is non-Abelian. That is, the transformations that need to be considered when doing the scrunching do depend on the order in which the processes are performed. That makes them much trickier to handle. However, one of the prizes of overcoming greater trickiness is greater richness. The appropriate non-Abelian local gauge transformations that are needed to show how indolence results in the weak force and at the same time to show that it stems from the same root as the electromagnetic force (so it becomes known as the 'electroweak force') has been identified and quite rightly earned its identifiers, Stephen Weinberg (b.1933) and Abdus Salam (1926–96), their Nobel Prize (in 1979).

The gauge invariance underlying gravitation remains to be found, so it is quite possible that my vision—a vision, or at least the hope, shared by many theoretical physicists—will remain unrequited.

* * *

Let me once again summarize where we have reached. In this chapter I have introduced what might be the origin of the forces that were originally classified as electric or magnetic, but by standing back (in a higher dimension) are revealed to be manifestations of a single force, electromagnetism. Various aspects of electricity and magnetism have emerged as we looked more closely at space, and used symmetry to discover that they can be regarded as arising from Nature's indifference to the modification of waves by scrunching, technically 'gauge invariance'. Thus, empty space turns out to be more intricate and subtle than it seems at first sight, and the uniformity that it inherited from its parent Nothingness is beyond vision, invisible, there, and responsible for the forces that bind and destroy, and in turn are responsible for the extraordinary workings of the world.

8

Measure for Measure

The origin of the fundamental constants

The fundamental constants, quantities like the speed of light ($c = 2.998 \times 10^8$ metres per second), Planck's constant ($h = 6.626 \times 10^{-34}$ joule-seconds), Boltzmann's constant ($k = 1.381 \times 10^{-23}$ joules per kelvin), and the fundamental electric charge ($e = 1.602 \times 10^{-19}$ coulombs), play an extraordinary role in the consequences of the laws of nature. The laws effectively issue orders about how to behave given various parameters such as mass and charge, and the fundamental constants determine the magnitudes of the resulting changes. For instance, the laws of nature we call special relativity imply that space and time become mingled the faster one travels; the speed of light establishes the extent of that mingling for a given speed of travel. The laws of electromagnetism imply that a charged particle is deflected by an electric field, and the fundamental charge determines the extent of that deviation for a given strength of field. The energy of an oscillator, like a mass on a spring or a pendulum, according to the laws of quantum mechanics, can step up a ladder of values, and Planck's constant tells us the separation of the rungs of this ladder: if it were zero there would be no gaps between the rungs and the energy of the oscillator could be increased continuously; that Planck's constant is so small (note the 10^{-34}) implies that the rungs are so close that we don't detect the separation in everyday pendulums and wobbling springs. But it is there.

There has also been a great deal of discussion about the serendipity of the values we currently have, for even tiny deviations from their values, it is argued, would have catastrophic consequences for the emergence of life, of consciousness, and of the ability to wonder why they have those seemingly benign values. With even slightly different values, stars might not form, or if they did form might burn their fuel so fast that there was no time for life to evolve, and so on.

To my mind there are two classes of fundamental constants: those that don't exist and those that do. As might be suspected, the values of the ones that don't exist are much easier to explain than the values of the ones that do. The former are essentially a consequence of mankind making, over the course of its intellectual history, sensible but fundamentally inappropriate choices about how things should be measured and reported (for instance, length in metres and time in seconds). The latter, the constants that really do exist in a fundamental way, and thus which are the truly fundamental constants, are coupling constants that summarize the strength of the interaction between entities, such as the strength of the force between electric charges, the strength of the interaction of an electric charge with an electromagnetic field, and the strength of the nuclear forces that bind elementary particles together and into the structures we call atomic nuclei. They also include the gravitational constant ($G = 6.673 \times 10^{-11}$ joule-metres per square kilogram) for specifying the strength of the gravitational field due to a massive body, and therefore which establishes the orbits of planets around their stars, contributes to the formation of galaxies, and determines the acceleration of an apple's fall.

Although tables of fundamental constants express them with units, such as the speed of light being so many metres per second, they shouldn't really have units. Put another way, the fundamental constants that don't exist all have the value 1 (so, $c = 1$, not $c = 2.998 \times 10^8$ metres per second), and the fundamental constants that do exist are best expressed in such a way that they too have no units. As I shall explain, instead of the fundamental charge having the value $e = 1.602 \times 10^{-19}$ coulomb, it is best expressed in a form that has the value 1/137. The other actual fundamental constants are all best expressed similarly as various other pure numbers. As will become clear, I think I can explain the value 1 but not values like 1/137. At present, we really have no idea where the numbers like 1/137 come from, and I shall make no pretence of knowing any better than anyone else. That is a shame, because it is these numbers that govern our existence and the emergence of thought: had 1/137 turned out to be 1/136 or 1/138 instead, we might not be here to know.

These remarks need to be elaborated so that you can see what I have in mind and why I think there are the two classes of constant. I shall not deal with all the fundamental constants (there are about a dozen important ones, and a number of combinations that are treated as though they are of the same rank). I shall just select a handful that I regard as truly fundamental and discuss their origins.

* * *

I'll begin with perhaps the most important fundamental constant of all, the speed of light, c (from *celeritas*). I consider it to have that rank because, even though it doesn't exist, it governs the structure of spacetime, the arena of all action.

There is more to space than meets the eye. Isaac Newton (1642–1726 old style), not to mention René Descartes (1596–1650) and that extraordinary mind of antiquity, Aristotle (384–322 BCE), who both inspired and suffocated thought, and we ourselves all glance at space and see that it is three dimensional. Albert Einstein (1879–1955), standing on the shoulders of others, changed all that. His special relativity (of 1905, his *annus mirabilis* but with more glory still to come as 'special' evolved into the even more extraordinary 'general') invites you to accept that space is entwined with time, and that what you thought of space and of time should not be regarded as each one separately but as components of a single arena, namely spacetime. That theory brings more discomfort and overthrow of the seemingly secure, for you then have to accept that what you regard as space and time are not what your neighbour might identify. If that neighbour is moving (most neighbours are, even if they are merely strolling, driving, or flashing past in a rocket), they have different perceptions of what component of spacetime is space and what is time.

It all depends on how fast you are moving. If you and I are not moving, then what you and I regard as space and time are exactly that: space and time. But suppose you are moving: you are walking, driving, or rocketing. Then you modify your perception in an extraordinary way: time rotates into space and space rotates into time. You are perfectly capable of asserting that something stationary is at a fixed location on a coordinate you regard as 'space'. But I, as a neighbour walking past, perceive space and time differently, and I no longer agree with your allocation of space and time to the event. The faster I travel relative to you the observer, the more my perception of time is rotated into my perception of space, and vice versa. Each of us, as we go about our daily activities, perceives space and time differently: your space is not mine, nor is your time mine (unless we are moving at exactly the same speed, including sitting still). The differences show up only if our relative speeds are very high, approaching the speed of light. That is merciful, for otherwise science and society would probably both be impossible. Nevertheless, it is the case that the fabric of reality is such that spacetime resolves differently for each of us and depends on our relative state of motion (hence 'relativity').

Now we can return to the role of the speed of light. It is sometimes thought of as a puzzle why there is a limiting speed for the propagation of information, with it being impossible according to special relativity to exceed *c*. Why is there such a limit? Could it be that there is a kind of viscous drag like that responsible for imposing a terminal velocity on a ball as it drops through a viscous medium? Is space viscous and the speed of light the terminal velocity for information as it drops through it? No: the explanation is much deeper and therefore simpler than that. The speed of light is simply the speed at which you have to travel for your perception of time to be rotated wholly into it appearing to be space. There is simply no further degree of rotation possible. There is no such thing as viscous drag on information passing through space: the limiting speed is a feature of our perception of space and time themselves.

But why does the speed of light have its particular value (of exactly 299 792 458 metres per second, about 670 million miles per hour)? The explanation lies in an artefact of human bureaucracy, boiling down to the fact that by convention we measure lengths in metres and not seconds. Various suggestions were made when the metre was first defined (in 1790, with French revolutionary zeal for the rationalization of just about everything, including the aristocracy). An early suggestion was to define it as one ten-millionth of the distance from the North Pole to the equator along a meridian, chosen diplomatically rather than sensibly to lie more or less half way between Paris and Washington in their respective young republics, and therefore which unhelpfully both started and ended in the sea. Compromise was then cast aside and the choice of meridian was shifted to the one that runs through Paris. A metal rod was then cast to enable this ultimate standard to be publicized and used more widely with greater convenience.

Since then, bearing in mind that the Earth effectively breathes and consequently the fiducial distance is not constant, and in principle therefore nor is the metre, a more precise and unchanging definition has been accepted. The metre is now defined as 1/299 792 458 of the distance that light travels (in a vacuum) in one second. Thus it is that all measurements of length are actually measurements of time. For instance, a person 1.7 metres tall could be recorded as being 1.7/299 792 458 seconds, or 5.7 nanoseconds, tall. Although light travels 299 792 458 metres in one second, that could be reported as 299 792 458/299 792 458 seconds, or 1 second. So what is the speed of light? If it travels through 1 second in 1 second, its speed (distance divided by time) is 1. No units; just 1. If you are in a car travelling at 100 kilometres per hour (that is about 28 metres per second), you should be able to work out that your speed is actually merely 0.000 000 093. At such a slow speed, it is obvious that you can ignore relativistic effects and be confident that your space hasn't been rotated into resembling time (compared to my assessment, as you flash past) and we should have no argument, at least to within reasonable precision, about whether two events are simultaneous.

I hope you can now accept that $c = 1$. That until now you thought it was expressed as a lot of metres per second turns out to be a historical accident: for perfectly understandable and sensible reasons, civil society measured distance and time in different units. Measure them in the same units, and a profoundly important fundamental constant effectively disappears. From now on, whenever I refer to a length *L* conventionally measured in metres, I shall denote it $L^{\dagger}$, the dagger signifying that the metres have been done to death and that henceforth it is to be reported in seconds. All speeds now lose their units and are pure numbers.

* * *

I suspect you might be wondering whether other rabbits can be pushed back into hats. What about probably the second most important fundamental constant, Planck's constant, *h*? Just as the speed of light essentially introduced relativity into science, Planck's constant effectively introduced quantum mechanics, so culturally they are of similar potency. The vanished *c* pervades all the formulations of special relativity; could it be the case that *h*, which is present in all formulations of quantum theory, should disappear too simply because a property historically was reported in convenient but fundamentally inappropriate units?

The German physicist Max Planck (1858–1947) was the initiator of quantum mechanics in what he regarded as act of despair. That despair was directed at the failure of classical physics, which he rightly loved, to account for what was thought to be the elementary problem of the colour of light emitted by an incandescent body, essentially why red hot became white hot as the temperature is raised. Classical physics had led to the erroneous conclusion that all objects should be white hot even when mildly warm. According to classical physics, there should be no darkness. Moreover, and worse, any object, even the merely warm, should devastate the countryside with gamma radiation. Planck's despair led him to suppose in 1900, or shortly before, that if something oscillated with a certain frequency, then it could exchange energy with the rest of the world only in packets, 'quanta', of energy, with the size of the packet proportional to the frequency: low-frequency oscillating things could exchange little packets; high-frequency oscillating things could exchange only big packets. Classical physics had supposed that an oscillator of any frequency could exchange energy in any amount; Planck's hypothesis supposed that energy is 'quantized', or exchanged in packets. That simple but revolutionary suggestion, which Planck seems to have hated as it was contrary to all his classical upbringing (Einstein had similar difficulties with quantum mechanics in general), accounted for the colour of hot objects, indeed the colour of objects at any temperature. We now know that it accounts for the colour of the Sun, at around 5772 K in the light-emitting surface regions, and of the entire universe, which has cooled to a miserly 2.7 K yet still glows with radiation characteristic of a body at that temperature.

In conventional physics, energy is reported in joules (J). A joule is quite a small unit, but very suitable for everyday discourse. For instance, each beat of a human heart requires about 1 joule of energy. The battery of a currently typical smartphone stores around 50 kilojoules of energy. The joule is quite a recent introduction, replacing a ragbag of earlier units that included calories, ergs, and 'British thermal units'. In the nineteenth century, as thermodynamics and the science of energy was emerging, heat was reported typically in calories and work was reported in ergs.

Here is an analogy to introduce an important point. There was once considerable interest in the efficiencies of steam engines, and therefore in the relation between calories of heat supplied and ergs of work produced. Elaborate experiments were performed to establish 'the mechanical equivalent of heat', the conversion factor, then perceived as a rather lowly fundamental constant, that could be used to convert the measurements of one form of energy into another. However, although those experiments were an important component of our intellectual progress, they were in another sense a complete waste of time. Had the early investigators measured heat and work in the same units, both calories or both ergs, then the conversion factor, that particular fundamental constant, would have been 1. That is now the case (except in a few isolated archaic islands of activity, including everyday food science) with the joule being used to report both forms of energy. The 'mechanical equivalent of heat' is now history or, to put it another way, 1.

I am sure you can see the parallels in this activity with the arguments I have been presenting about the real fundamental constants, or at least the ones that don't, or shouldn't, exist: choose the same units for related quantities, and conversion factors become 1. Planck's constant is a candidate for this treatment. It was introduced to relate the frequency of oscillation to the size of the corresponding energy packets, the minimum size of the quanta that can be transferred.[1]

The way ahead should now be clear. Let's do away with joules and report energy as a frequency, in cycles per second. Whenever I want to report energy as a frequency, I shall denote it $E^{\dagger}$ and report it as so many cycles per second. There is no longer any need for a conversion factor between them, any more than there is a need to report and list the mechanical equivalent of heat or, having decided to report distance in seconds, for there to report and list the speed of light. Planck's constant has become 1. Joules, like calories and ergs, are now history. There might at first thought seem to be profound implications for quantum mechanics if $h = 1$ rather than its conventional tiny value: but that is not the case, as I shall develop after swilling out a few more items littering the gutters of the Augean stable of conventional units.

A semi-final point in this connection is that with the shenanigans developed so far, Einstein's formula $E = mc^2$ becomes $E^{\dagger} = m^{\dagger}$ with both properties reported as frequencies. You are welcome to keep the form $E^{\dagger} = m^{\dagger}c^2$, but to do so you now have to accept that $c = 1$, as I have already argued. A truly final point is that you can now see that because $E^{\dagger} = m^{\dagger}$, energy and mass are the same.

* * *

Just about everyone these days (except in the USA, in company with Burma and Liberia) expresses mass in kilograms and its fractions (grams) or multiples (tonnes, 1000 kilograms). The kilogram was originally defined (back in the 1790s) as the mass of a litre of water at a certain temperature. Like the metre, that definition was refined and replaced by a standard kilogram, the 'International prototype of the kilogram' (IPK), a cylinder of platinum-indium alloy maintained at the *Bureau International des Poids et Mesures* in Sèvres, on the outskirts of Paris, and with various secondary copies spread around the world. Unfortunately, even the IPK is not perfectly stable, for impurities evaporate from it, air diffuses into it, and minute scratches are caused when it is handled, so what is meant by 'a kilogram' is slowly changing. The current proposals are to define the kilogram in terms of Planck's constant, an eternal constant (as far as we know), so that the meaning of 'a kilogram' is fixed for all time and anyone with access to the fundamental constants knows exactly what is meant. What does that mean for our current purposes?

Let's adopt the view that humanity, in its usual muddly way, made a collective but sensible mistake when it adopted the kilogram as a measure of mass. Instead of the kilogram, suppose it had adopted the second, or more precisely the 'cycles per second', just like a frequency. With extraordinary prescience, it could have done that by reporting not m but $m^{\dagger} = mc^2/h$ and reporting mass in oscillations per second. A mass of 1 kilogram, for instance, would then be reported as 1.4×10^{50} cycles per second. If you think of yourself as a well-proportioned 70-kilogram person, from now on you should think of your mass as a breathtaking 9.5×10^{51} cycles per second by converting mass in kilograms into an energy in joules by multiplying by the square of the speed of light (that is, use $mc^2 = E$), and then using Planck's constant to express that energy as a frequency in cycles per second. The unit 'cycles per second' is becoming a little tedious to write and to read; it is actually the definition of the unit 'hertz' (Hz), which is named after the regrettably short-lived pioneer of radio communication, Heinrich Hertz (1857–94), so 1 cycle per second is 1 hertz. By adopting this procedure of multiplying by c^2 and then dividing by h, your mass will turn out to be around 9.5×10^{51} hertz. That might seem a silly way to report mass, but that isn't the point. In everyday practice the kilogram is sensible and useful. I, though, am trying to get to the root of reporting data in the most consistent manner and in the processes bringing my typographical dagger to the throats of conventional units.

* * *

We can now see why setting $h = 1$ is of no consequence in the physical world, in the sense that it leaves quantum mechanics intact. One way to do that is to show that the Schrödinger equation (which I introduced in Chapter 3 as one of the principal components of quantum mechanics) remains unchanged apart from the interpretation of its symbols, but equations as complicated as his are confined to lurk in the shadows of this book, the Notes.[2] Another way is to lead you into the foundations of his equations. That turns out to be possible, for foundations, even in science, are invariably simpler than the edifices they support.

If you are a commuter, you are already half way to understanding quantum mechanics. The term 'commuter' stems from the common practice of selling a ticket for 'there and back' for less than the sum of the individual 'there' and 'back' fares: the return fare is 'commuted' (from the Latin *commutare*, 'to change, to alter'). Put another way, the cost of the 'back' fare is not the same as the 'there' fare (provided you have already invested in the 'there' fare). Quantum mechanics differs from classical mechanics in much the same way. The analogy is as follows. The fare for travel 'there' becomes multiplication of linear momentum by position; the 'back' fare becomes multiplication of position by linear momentum (note the opposite order). The two 'fares' are not the same, and the difference is called the 'commutator' of the position and linear momentum.

A railway company can adjust its fares for commuters at whim. Nature appears to have settled on a particular standard commutation, and the reduction in the round trip is equal to a minor (but far-reaching) modification of Planck's constant.[3] That is, going 'there' with linear momentum multiplied by position minus coming 'back' with position multiplied by linear momentum, is proportional to h. The whole of the deviation of the predictions of quantum mechanics from classical mechanics springs from that commutation of the 'there and back' fare, and all quantitative aspects stem from the fact that Nature's board of directors have allowed the commuter discount to be proportional to Planck's constant.

In conventional units Planck's constant is so tiny (but pregnant) that the board of directors of classical mechanics decided it wasn't worth the administrative hassle of giving the commuters any discount. It is easy to see their point. It would be like getting a 1 penny discount on a fare of many trillions of pounds. From that perfectly reasonable decision, classical mechanics emerges.

Reasonable it might be, but wrong it is too. The actual board of directors of Nature insist on maintaining the discount. The most successful ever mathematical description of matter and radiation, quantum mechanics, differs from classical mechanics by the simple offer of a discount to commuters, yet has consequences of the profoundest implication. As I have indicated, Newton and his contemporaries and immediate successors had no inkling of the lack of commutation of position and momentum, and developed his and their wonderful cathedral of theoretical structure we call classical mechanics on this oversight. From it grew an understanding of the heavens, for who cares about such a tiny discount when bodies as big as planets are encircling the Sun? But when scientists turned their attention to electrons in atoms, when the 'there' and 'back' fares are themselves very tiny, then the commutation discount is tremendously significant. On a fare of one pound, the discount could be as much as 50 pence; it simply cannot be ignored.

How, then, is it possible to set *h* equal to 1 rather than to a miniscule 10^{-34} and still end up with classical mechanics being appropriate for everyday objects? Wouldn't that mean that any everyday position and momentum qualifies for a significant commuter's discount? The weasel position is that I finessed the problem by leaving off the units of 10^{-34}. The values of everyday positions and momenta, which might have quite ordinary everyday values in the old units of position in metres, mass in kilograms, and speed in metres per second, become enormous when expressed in the new units of position in seconds, mass as a frequency in cycles per second, and speed in no units at all. As a result, the product of position and momentum for an everyday object also becomes enormous in the new units, and far, far bigger than 1.[4] In the old way of looking at things, position and momentum had everyday values and h was exceedingly small. In the new way, it is h that has an everyday value (of 1) and position and momentum are exceedingly large. The result, the discount being negligible, is effectively the same and the consequence of that negligibility is the same too: you don't need quantum mechanics for everyday objects.

I need to mention here that great clarifier of human thought, Heisenberg's uncertainty principle, which he formulated in 1927, for it stems from the lack of commutation of momentum and position. The principle states that it is not possible to know, with arbitrary precision, position and momentum simultaneously. Quantum mechanics, much to the discomfort of those brought up in the classical tradition (I include Bohr and Einstein), thereby reveals that we have to make a choice when seeking to specify the state of a system. It instructs us to choose a description in terms of position or choose a description in terms of momenta, either of which can be specified with arbitrary precision. If you insist, as a result of your classical conditioning, to speak in terms of both descriptions, believing that only then can your description of the world be complete, you are brought up short by the uncertainty principle, which implies that the two descriptions are intrinsically incompatible. If you cannot shake off your conditioning as a classical physicist, you are led to the view that quantum mechanics disallows a complete description of Nature. A much more positive view, however, is that what the practitioners of classical mechanics took to be 'complete' was in fact unattainably *over*-complete. Quantum mechanics tells us that the use of both descriptions simultaneously is inconsistent. It is a bit like starting a sentence in one language and ending it in another. You have to choose your language, for otherwise your message will be incomprehensible and your interlocutor, in this case the universe, will look at you blankly. Quantum mechanics strips away this common-sense inspired error and accepts that completeness exists in one language or the other, in position or momentum, not both. When that is accepted, the description of the universe is simplified (but still not simple). That is why I regard the uncertainty principle as a great clarifier.

* * *

I have done away with *c* and with *h*, the hinges on which relativity and quantum mechanics swing. Is there room in the graveyard for other fundamental constants? If I were to identify the single most important fundamental constant that is in effect the hinge of thermodynamics, then I would choose Boltzmann's constant, ***k***. It occurs in the all-important Boltzmann distribution that I lauded in Chapter 5, it is carved into Boltzmann's tombstone for his definition of entropy, and it occurs in disguise (among other entities, as the gas constant in the discussion of gases) subversively throughout thermodynamics. It is, however, completely unnecessary and can be eliminated and buried by using arguments not unlike those that I have used to do away with *c* and *h*.

The mistake, once again a sensible, understandable, and laudable mistake, goes back to Celsius and Fahrenheit, whom I introduced in Chapter 4 as the inventors of early temperature scales, and was compounded by Kelvin's introduction of a seemingly more natural absolute scale. First, you need to recognize that all three were seduced by convention, with perhaps Celsius the least seduced. In our current world, the hotter the object the higher the temperature, on all three scales. As I have mentioned, Celsius originally had it going the opposite way, the hotter the lower on his original scale. I think he was, unknowingly, on the right track, for in a variety of ways I think 'the hotter the lower' is more natural at a thermodynamically fundamental level, as I shall explain. But all three, in my view, got it wrong by introducing a new unit of measurement (the degree, and later the kelvin, K) to report temperature, just as introducing metres to measure length instead of using seconds also resulted in unnecessary confusion that became apparent as science matured. In the latter connection you have seen that had length been measured in seconds, then there would have been no need to introduce the fundamental constant *c*, the speed of light. In a similar vein, I shall argue that had temperature been reported in the same units as energy, then there would have been no need to introduce Boltzmann's constant.

There are obviously a number of matters that I need to explain. Boltzmann's constant, which is so many joules per kelvin, can be regarded as a way of converting kelvins to joules. If you have already agreed to report temperature in joules, then there would be no need to convert it into those units. Moreover, if there is a uniform relation between temperature in kelvins and joules, then there is no ambiguity in the change of units. You might end up with some unfamiliar funny numbers, but unfamiliar funniness is not one of the criteria of acceptability in science (although it might be one in the pragmatic everyday world). For example, with the currently accepted value of Boltzmann's constant, a mild 20 °C (293 K) would be reported as an unfamiliar 4.0 zeptojoules (zepto is the perhaps unfamiliar but useful prefix denoting 10^{-21}) and water would boil at 5.2 zeptojoules.

If you agree to report temperatures in joules (or its submultiples, such as zeptojoules), then the gradations on our thermometers will have to be in joules or a submultiple of joules, and each degree on the current Celsius scale becomes 0.0138 zeptojoules. Once you have done that, there is no need ever again to invoke Boltzmann's constant in any expression. In effect, if you insist on using the equations you come across in current textbooks, then wherever *k* appears you should ascribe to it the value 1. Now *k* has gone the way of *c* and *h*. It is a superfluous fundamental constant that emerged simply because the early scientists were misled by sensible everyday practice into introducing a new but unnecessary unit for the measurement of temperature.[5]

But what did I mean by Celsius originally being less in error than Fahrenheit and Kelvin, and it being better to think of temperatures unfamiliarly going down as things get hotter? Here I have in mind the fact that many expressions in thermodynamics, and in particular its cousin 'statistical thermodynamics', which provides the link between the molecular and the bulk, between the individual and the crowd, are strikingly simpler if expressed as the *inverse* of temperature (that is, as $1/T$ rather than T, not simply reversed with 0 and 100 changing places). The mathematics seems to be crying out to us that a natural temperature scale is one in which the scale should be not simply reversed but turned upside down. With temperature already in zeptojoules, its inverse would be reported in 'per zeptojoules'. In this way (I leave the little arithmetic to you), the boiling point of water would be 0.19 per zeptojoule and its freezing point would be higher at 0.27 per zeptojoule.

From now on, I'll express all temperatures upside down and converted, as so many 'per zeptojoules', and denote the newly defined temperature by the letter Ŧ (tee-dash). As I have forbidden myself to quote any formulas except in the safe space of the Notes, to which I refer you,[6] you will have to accept my word that if you take any formula in statistical thermodynamics, then it looks, and is, simpler when T is replace by $Ŧ$. But there is more to that replacement than appearance.

Everyone (well, almost everyone) knows that you can't reach the absolute zero of temperature. The third law of thermodynamics expresses that unattainability in more sophisticated, scientifically acceptable terms, adding 'in a finite number of steps' and a bit more, but that is its general gist. It might seem odd that $T = 0$, the bottom of the Kelvin scale, can't be reached in a finite number of steps. But $T = 0$ corresponds to $Ŧ = \infty$, and there is probably little psychological rejection of the impossibility of reaching infinite $Ŧ$ in a finite number of steps.

A deeper simplification comes from the exploration of various equations of statistical thermodynamics. Although negative absolute temperatures (temperatures like –100 K) are meaningless in ordinary thermodynamics (they are like negative lengths: something can't be –1 metre long), there is nothing wrong with fiddling about with the equations of statistical thermodynamics and seeing what happens to various properties (for instance entropy) as the temperature is allowed to sink though zero and become negative, and even become negatively infinite. For instance, you could take any of the formulae in note 6 and see what happens when you insert a negative value of the temperature. Typically, nasty things happen when you do that, with the properties showing sharp jumps or squirting off to infinity as the temperature passes though zero. However, if the same properties are plotted against $Ŧ$, then all these jumps and squirts disappear, and all the properties behave smoothly. This taming of the properties strongly suggests (it is no more than that) that *Ŧ* is a more fundamental measure of temperature than T. But, I shall now argue that it is not quite fundamental enough: it hasn't reached rock bottom in fundamentality.

I am sure that you are seeing a pattern emerging through these chapters, with everything becoming simplified by being expressed either in seconds (time and distance) or as a frequency in 'per second' (energy). You have also seen that inverse temperature *Ŧ* is an inverse energy in 'per zeptojoules'. Now note that we can convert that inverse energy to inverse 'per second', which is simply seconds.[7] Then 20 °C becomes 0.16 picoseconds (pico is the prefix denoting 10^{-12}), water freezes at 0.18 picoseconds and boils at 0.13 picoseconds.

At this stage, the three fundamental constants of relativity, quantum mechanics, and thermodynamics, c, h, and k, have become redundant. Put another way, if you insist on using equations in which they appear (such as $E = mc^2$), and have chosen to express the properties (such as E and m) in related units (such as seconds or their variations), then you have to set each fundamental constant equal to 1 and there is no longer a mystery about their origin.[8]

* * *

I now leave these non-existent fundamental constants which I can explain and turn to the ones that really exist and I cannot explain. There are just two that I shall mention, but others are lurking in this Pandora's box of the currently inexplicable. Both are coupling constants, governing the strengths of two varieties of interaction.

I have already mentioned the fundamental charge, e, which expresses the strength of electromagnetic interactions, such as the strength of the attraction between two charges and the strength of the interaction of an electron (which has charge $-e$) with an electric field, such as that in a radio wave. The size of this fundamental constant affects the strength of the interaction between electrons and nuclei in atoms, and therefore the sizes and properties of atoms, the strength of the bonds between atoms and therefore the formation of compounds, and the strength of the interactions of electrons in atoms and molecules and the electromagnetic field, so it also affects the colours of materials and the intensity of those colours. It plays a role within atomic nuclei, for the positively charged protons within nuclei are subject to intense mutual repulsions.

Once again it is best to detach the magnitude of the fundamental charge from human-inspired units and to express it as a pure number. Whenever you see units attached to a constant, you can't be sure that it is large or small: large or small compared to what? In its case, the fundamental charge is commonly wrapped in other fundamental constants to produce a dimensionless number, the 'fine-structure constant', α (alpha), which is so-called because it was introduced to explain some of the detailed structure of the spectrum of hydrogen atoms. It has the value I mentioned earlier, namely 1/137.[9] That α is so small reflects the weakness of electromagnetic interactions (compared to the strong force at work within nuclei) and is responsible for molecules, which are held together by electromagnetic interactions, being much more malleable than nuclei in the sense that they can be torn apart and reassembled in chemical reactions. If α were close to 1, there would be no chemistry, molecules, if they existed at all, would be the size of atomic nuclei, and life (a highly elaborate chemical reaction) would not have emerged. The universe would have been biologically silent.

No one yet knows why α has the value 1/137. In one scenario, all the forces once had the same strength but as the universe expanded and cooled their strengths diverged, and 1/137 emerged as the strength of one of them. That value, I presume, will be explained once a more comprehensive theory of the inception, structure, and evolution of the universe has been formulated, but at present its value is a mystery. That is not to say that a variety of concoctions of numbers like π and $\sqrt{2}$ have not been cobbled together, some with impressively close values to the experimental value.[10] However, they are cobblings together with no reliable theoretical foundation and none of them has been accepted by the scientific community as anything other than numerological jugglings. The problem, though, is of enormous importance for understanding the universe and our place in it. There are similar coupling constants for the strong and weak forces that play a role in nuclear structure. Some future theory of the fundamental forces (and the fundamental particles they act on) will have to account for all their values.

The only other coupling constant I shall mention is the one that governs the strength of gravity. This constant, the 'gravitational constant', G, appears in the inverse square law of gravitational attraction between two masses.[11] The gravitational constant can be turned into a dimensionless quantity, α_G, analogous to the fine-structure constant, effectively replacing the square of the charge of an electron (which appears in α) by the square of the mass of an electron, when it turns out to be 1.752×10^{-45}.[12] Now you can see that it is a very tiny quantity, and conclude that gravity is a far, far weaker force than electromagnetism. That is beneficial for the emergence of thinking entities, currently at least us, for it gives time for star formation, galaxy formation, the persistence of planets in orbits around their stars, and the inception and evolution of folk. If it were much stronger, we—everything—would all be in a big black hole together (and not know it).

No one has a clue about the origin of the value of *G*. Current speculations include the possibility that it was once strong but fizzled out to nearly nothing when the universe cooled (much like the fine-structure constant, but its fizzling out went further). Some speculate that it really is strong still, but that much of the strength of gravity has leaked out into the six or seven dimensions that have yet to unfurl and be detectable. No one knows why gravity is so weak, and certainly not why α_G has its current value, and I will not pretend otherwise.

* * *

Where are we? The laws of nature control the behaviour of entities in a general way, but their quantitative consequences depend on the values of various fundamental constants. These include the speed of light, which is central to relativity, Planck's constant, which is central to quantum mechanics, and Boltzmann's constant, which is central to thermodynamics. However, I have tried to show that if all physical observables are expressed in the same or related units rather than being trapped in a pragmatic but motley collection of human-devised units, then these three fundamental constants can be discarded. Put another way, if you go on insisting that they appear in equations, then you can set them all equal to 1 provided you express all observable properties in related units (I have chosen seconds and its variations). There is another class of fundamental constant which consists of effectively coupling constants that express the strengths of various forces, such as the electromagnetic force and the gravitational force. No one yet has a clue about why they have their current, and for us serendipitous, values.

9

The Cry from the Depths

Why mathematics works

Many laws of nature are expressed mathematically; all of them, even those that are not intrinsically mathematical (like whatever laws might be formulated to describe evolution by natural selection), acquire greater power when developed mathematically. One of the first scientists to consider this point was the influential Hungarian mathematician Eugene Wigner (Wigner Jenö Pál, 1902–95), who raised the question in a lecture on 'The unreasonable effectiveness of mathematics in the natural sciences' in 1959.[1] His perhaps wisely timid conclusion was that its unreasonable effectiveness is a mystery too deep to be resolved by human reflection. Others have added to the general sense of despair that of current mysteries, this one is likely to endure.

An alternative, more positive view in contrast to Wigner's cautious pessimism, is that the effectiveness of mathematics is not unreasonable and, instead of being perplexing, offers an important window opening on to the deep structure of the universe. Mathematics might be the universe struggling to speak to us in our common language. In the course of this chapter, I shall try to remove the tinge of mysticism, heaven forbid, that might seem to infect that remark.[2] The existence of mathematical versions of the laws of nature perhaps points to a serious question, and let's hope a rewarding answer, concerning what might be the deep structure of the fabric of reality. Perhaps it points to the deepest question and for aeons the most perplexing and compelling question of all: how there has come to be what there is.

* * *

It is undeniable that mathematics is an extraordinarily potent and successful language for conversing with the universe. At the most pragmatic level, an equation that summarizes a physical law can be used to predict a numerical outcome, as in the prediction of the period of a pendulum from its length. Just look at the astonishing ability of astronomers to predict the orbits of planets, the incidence of eclipses, and (today as I write) the appearance of a supermoon, the coincidence of a full moon and the moon's close approach to Earth. Then there are the unexpected consequences that emerge from a mathematically stated law that are verified by observation. Among the most famous of these is the prediction of black holes by listening to the content of Einstein's general theory of relativity, his theory of gravitation. It has been said, ironically of course, that no experimental observation can be accepted unless it is supported by a mathematically formulated theory. World economies have bloomed, and sometimes withered, under the impact of the pursuit of mathematical formulations of the laws of nature. A very high proportion of the industrial output of nations has been ascribed to the implementation of quantum mechanics and its mathematical formulation.

There are, of course, aspects of our understanding the universe and our physicalization of it that are not expressed mathematically. Right at the beginning of this book, and in passing a moment ago,

I drew attention to one of the most far-reaching theories of the universe, the theory of natural selection as an explanation of evolution. That theory is not intrinsically mathematical in the sense that it is not expressed by a formula, yet is of great potency, perhaps applying throughout the universe wherever there is anything that can be regarded as 'life'. It has even been applied to the emergence not just of new species but of whole new universes. It can be expressed as a kind of law of nature, with Herbert Spencer's 'survival of the fittest' being a crude but pungent approximation. But when developed mathematically, for instance by modelling population dynamics, as I shall mention again in a moment, the qualitative version of the theory becomes immeasurably and quantitatively enriched in the sense that it can make quantitative predictions.

Biology as a whole is perhaps a less obvious domain of mathematical exposition. This branch of human knowledge was largely nature walks until 1953, when Watson and Crick established the structure of DNA and almost at a stroke rendered biology a part of chemistry and therefore a member of the physical sciences and all the power that that implies. Nevertheless, it is hard to point to specifically mathematical biological laws, except (coming back to DNA) the laws, including the encoding, of inheritance. But there are numerous and varied candidates for illustrating the direct role of mathematics in biology. They include the analysis of populations of predators with access to prey and in a certain sense the similar business of devising harvesting and fishing strategies. Periodic phenomena of all kinds are typical of organisms, as a moment of reflection about ourselves, breathing and with beating hearts, and our slower circadic rhythms, will confirm, and such oscillations are open to mathematical description. Likewise, waves of difference, such as difference of numbers of people infected and not infected in an epidemic, and waves of electrical potential difference as in the propagation of signals along nerves as we think and act, and waves of muscle activity as a fish flexes itself autonomically (even when decapitated) in transverse travelling waves to propel itself through water, are all aspects of biology that can be treated mathematically.

The brilliant and tragically maligned Alan Turing (1912–54) was perhaps the first to undermine the reputedly incredibly ugly Aesop (possibly 629–565 BCE, if indeed he ever existed), and show that the mathematical treatment of waves of spreading chemicals through containers of various shapes, shapes like leopards for instance, accounts for the patterns of animal pelts, including how the leopard got its spots, the zebra its stripes, and the giraffe its blotches, and the intricate beauties of butterfly wings. Even the elephant got its trunk from a wave of chemicals spreading through its early embryo in accord with mathematical laws expressed as equations and their solutions.[3]

Sociology, that elaboration of biology applied to human populations sometimes modelled as rats, emerged in the late eighteenth century; the word was coined by Emmanuel-Joseph Sieyès (1748–1836) in 1780, but the subject came to fruition in the late nineteenth century and acquired its mathematical structure in the twentieth as elaborate statistical models could be explored numerically on computers. Although its early impulse was to identify the laws of human behaviour, its principal achievements have been the development of statistical methods for analysing and sometimes predicting the most probable or average behaviours of populations of individuals. Such statistical modelling is essential for the effective running and governance of societies, but no fundamental laws, other than the laws intrinsic to statistics itself (such as bell-shaped distributions of random variables) have emerged, despite the longing to identify them.

Theology, the study of the intrinsically elusive and incomprehensible divine, the scholarly version of searching for the grin of the Cheshire cat, has no need of mathematics. Nor, of course, do those other vastly more positive creations of working brains, such as poetry, the arts, and literature, which so enhance the mundane with engaging and sometimes appalling fantasies. Statistics is an exception, for it helps, for instance, to untangle Marlowe from Shakespeare. Music lies perhaps on the borderline and might be a point of entry into a science of aesthetics where mathematical insights could prove invaluable by examining chords and sequences of notes in relation to possible resonance circuits in the brain.

I now need to deflate this account to a degree. Despite all these varied applications of mathematics, they are not in themselves laws. Apart from the numerical analysis of data that statistics pursues, in every case (I think) the mathematical component consists of the analysis of a model. That is not the stuff of the fundamental laws of nature, it is the formulation of a complex arrangement of underlying fundamental physical laws. These are not even outlaws, they are sorties of organized gangs of outlaws.

* * *

At the simplest and most obvious level, mathematics works because it provides a dispassionate and highly rational way of unfolding the consequences of an equation that expresses a law in symbolic form. Thus, it is impossible to make reliable predictions from a non-mathematical statement, such as 'the fittest survive', and to predict, for instance, that primitive combinations of elements will in due course evolve into elephants. In contrast, reliable predictions can be made from a mathematical statement, such as Hooke's law that the restoring force is proportional to the displacement (the verbalization of the equation $F = -k_f x$): the period of a pendulum can be predicted accurately from its length.

I hear you cry 'chaos'. It is certainly the case that the unfolding of certain systems appears to be unpredictable, but that unpredictability must be interpreted with caution. A simple case of a system that shows chaotic behaviour is a 'double pendulum', in which one pendulum hangs from the bottom of another, and both swing in accord with Hooke's law. In this case the equations of motion of the pendulums can be solved and, provided the initial angles at which they are both pulled back are known exactly, the future angles at any time can be predicted exactly. The key phrase here is 'provided the initial angles at which they are both pulled back are known exactly', for even an infinitesimal imprecision in the starting angles results in wildly different subsequent behaviour. A chaotic system is not one that behaves erratically: it is one with a very high sensitivity to the starting conditions such that, *for all practical purposes*, its subsequent behaviour is unpredictable. Perfect knowledge of the starting position results (in the absence of externally intrusive effects such as friction and air resistance) in perfectly predictable behaviour.

One consequence of this intrinsic *practical* impossibility of matching prediction to observation is a shift in the meaning of experimental verifiability in science. It has long been held that a cornerstone of the scientific method is the process of comparing prediction with observation and revising the theory in the light of failure. But now we see that reliable prediction is not always possible, so has the cornerstone been sapped? Not at all. The 'global' prediction that the model leads to chaotic behaviour can be verified by testing the system against different starting conditions, and indeed the 'chaos' has certain predictable qualities that can also be verified. It is not necessary for the precise trajectory of a double pendulum to be predicted and verified for us to claim that the system has been understood and its behaviour verified. The laws of nature, in this case the gang of outlaws, would have been verified even in this case of quantitative unpredictability.

The human brain is a concatenation of processes far more complex than the mechanical triviality of a double pendulum, and it is therefore hardly surprising that its output—an action or an opinion, even a work of art—cannot and presumably will not ever be predictable from a given input, such as a glance or passing phrase. Theologians call this unpredictability 'free will'. As for a double pendulum, but on a far more complicated scale, we could claim to understand the workings of a brain, whether artificial or natural, in terms of the network of processes going on within it even though we had failed to predict the opinion it might have expressed, poem it had written, or act of slaughter it had initiated. The occurrence of 'free will' will therefore in a sense be a confirmation that we understand the working of a brain, just as the occurrence of chaos is a confirmation that we understand the working of a double pendulum. It is probably too much to hope that just as patterns of chaos are predictable for simple systems, so patterns of free will might one day be discovered. Perhaps, through psychiatry, they already have, but have not yet been formulated with precision.

* * *

The dispassionate rationality of mathematics might be all there is to its unreasonable effectiveness. Its effectiveness is perhaps not unreasonable: perhaps it lies in its reasoning and it being the apotheosis of rationality. The reason why mathematics works might simply be its emphasis on systematic procedure: start with the proposition of a model, set up a few equations that relate to its properties, and then unfold the consequences using the tried-and-tested tools of mathematical deduction. That might be all. But could there be more?

There are certain other pointers that the world might be mathematical in a deeper sense. My starting point here is the remark made by the German mathematician Leopold Kronecker (1823–91), who said that '*Die ganzen Zahlen hat der liebe Gott gemacht, alles andere ist Menschenwerk*' ('God made the integers, everything else is the work of man'). Thus all the wonderful achievements of mathematics are manipulations of entities, the integers, into patterns for which they were not originally intended, which was just plain old-fashioned counting. But where did the integers come from, discounting God's munificence as too easy an answer?

The integers can spring from absolutely nothing. The procedure belongs to that most etiolated region of mathematics known as 'set theory', which deals with collections of things without paying much or any attention to what the things are.

If you haven't got anything, you have got what is called the 'empty set', denoted Ø. I'll refer to this as 0. Suppose you have got a set that includes the empty set, denoted {Ø}. You have now got something, which I'll call 1. You can probably see where this is going. You might also have a set that includes not only the empty set but the set that includes the empty set. This set is denoted {Ø,{Ø}}, and because it has two members I'll call it 2. You can probably see now that 3 is {Ø,{Ø},{Ø,{Ø}}}, and contains the empty set, the set that contains the empty set, and the set that contains both the empty set and the set that contains the empty set. I won't burden you with 4, let alone anything more elaborate, because the procedure should by now be clear. What it achieves, of course, is the generation of the integers out of absolutely nothing (the empty set). And once you have got the integers, and force them to jump through all kinds of hoops, as Kronecker said, you end up with mathematics.

The analogy with the emergence of the universe from absolutely nothing should now be obvious, with Nothing somehow identified with the empty set, {Ø}. But it might be merely an engaging analogy and have nothing to do with the emergence of the universe, mathematical or not, from Nothing. Then again, it might be a profound insight into how there can seem to be something here and why mathematics is so successful as a language of its description and elucidation.

I can see that there are several problems with the analogy. They include the absence of rules about how the integers are linked into the structures we know as 'mathematical'. Having a list of integers is hardly worthy of the name 'universe'. Here the answer might lie in the axioms that have been proposed as a basis of arithmetic. Among these are the celebrated axioms proposed by the Italian mathematician Giuseppe Peano (1858–1932).[4] Once you have got arithmetic, you have got a lot of other stuff, because there is a celebrated theorem due to the German Leopold Löwenheim (1878–1957) and the Norwegian Thoralf Skolem (1887–1963) which implies that any axiomatic system is equivalent to arithmetic.[5] So, for instance, if you have a theory encompassing all the laws of nature that is based on a set of assertions (axioms), then it is logically equivalent to arithmetic, and any statements about arithmetic apply to it too. A wild speculation might therefore be that logical relations akin to those proposed in Peano's axioms were stumbled into and gave stability to the entity that emerged from nothing and which we call the universe. I am clearly groping in the dark here for meaning, and any reliable interpretation of this vision, if that ever emerges, will have to await deep advances in the understanding and elucidation of our cosmic roots. For the time being, these musings are but whimsy.

* * *

One big question, of course, is what do we mean by the universe being mathematics? What am I touching if it is only arithmetic? What do I see through my window if it is but algebra? Is my consciousness just a collaboration of integers dancing to the music of axioms? Is causality akin to, or actually, the unfolding of the proof of a theorem?

Take touch. Are we in some sense touching the square root of 2 or even pi itself? I can perhaps help you to see that you are. If we set aside the neurophysical aspects of touch that go on inside us when we make contact with an external object (and I am aware that you might say: 'But that is the entire point of touch, our mental response to it!'; bear with me), then touch boils down to the impenetrability of the touched to the toucher. Impenetrability is some kind of exclusion from a region of space, and as such we can understand the origin of the signal that carries 'touch' to the brain or into a neural reflex circuit that triggers withdrawal from possible danger or that furtherance of touch: injury.

The exclusion of one object from another stems from a very important principle propounded by the Austrian-born theoretical physicist Wolfgang Pauli (1900–58; yet another brief flame) in 1925 and generalized in 1940, earning him the 1945 Nobel Prize in Physics. This principle, which is intrinsic to quantum mechanics, concerns the mathematical description of electrons (among certain other fundamental particles), and asserts how the description must change when the names ascribed to two electrons are interchanged.[6] The consequence of the principle is that the electron clouds of two atoms cannot intermingle: one is excluded from the region occupied by the other. Touch has emerged from a fundamental principle of nature. Although I accept that this vision of touch still doesn't quite get to the heart of the business of what it means to touch mathematics, I hope you can accept that it is a step towards it.

Hearing is a form of touch. In this case the sensitive receptor is inside the ear, and the touch on it is that of molecules in the air that constitute the pressure wave and the impacts that they provide on the membrane of the drum. That the detector passes on the detection of these touches to a different region of the brain is why we think of hearing as distinct from touch; but fundamentally it isn't. Vision is also touch, but of a more subtle, buried kind. In its case the touch is between the optical receptor molecules in the rods and cones of the retina. They sit embedded in a cup-like protein until a photon of light excites them into a different shape. The protein can no longer accommodate them—touch again—and they pop out, allowing the protein to change shape slightly and trigger an impulse into yet another region of the brain where it is interpreted as a component of vision. Smell and taste too are aspects of touch—this time (so it is currently thought, although the mechanism is still controversial) molecules sniffed into the nose or alighting on the tongue touch its receptors and trigger a signal to still another part of the brain. All sensation is ultimately touch, and all touch is a manifestation of Pauli's principle concerning the mathematical nature of the world.

I have to admit, as I have already half admitted, that this account of sensation being a manifestation of a piece of mathematics is unlikely to be convincing, and nor have I dared pursue the triggering into the dark mysteries of the brain and the way it converts sensation into consciousness. How can it be convincing before we really know the deep nature of matter? I hope, though, that it is at least suggestive of our being intimately in contact ultimately with the integers and their elaborate organization into reality.

* * *

There is a final important matter, which could be a matter of life and death. Where does Gödel's theorem stand? Gödel's theorem, which was proved in a remarkable *tour de force* by the Austrian-born eponymous Kurt Gödel (1908–78, who famously starved himself to death in Princeton through fear of being poisoned) in 1931, essentially asserts that the self-consistency of a set of axioms cannot be proved from within those axioms.[7] Given that the laws of nature are mathematical, could it be that they are not self-consistent? Is my account of them systemically doomed? If the universe is one giant piece of mathematics, could it too perhaps not be self-consistent. Might it collapse under the weight of its own inconsistency?

There are escape hatches from this scenario. Gödel based his proof on a particular formulation of arithmetic, one version of which I specified in note 4. Suppose you ditch one of those statements, for instance the specification of what you mean by multiplication. Then that knocks a leg from beneath Gödel's proof and it fails. Arithmetic without × might seem a bit weird, but so perhaps did the version of arithmetic I mentioned in Chapter 8 when the outcome of 2 × 3 was not the same as the outcome of 3 × 2, yet it proved to be the key to understanding the physical world. Take out multiplication from arithmetic, Gödel is left stranded impotent on the wayside, and arithmetic turns out to be complete.[8] Who knows what the picture would be if furthermore 2 + 3 was not taken to be the same as 3 + 2? The bottom line, though, is that it is far from clear whether the conditions on which Gödel based his proof are applicable to the physical world (the only world), so pessimism is unfounded, the laws of nature might well be self-consistent, provably so, and there is no underlying logical fault line in the universe that might, in an instant, spread catastrophically and wipe us and all there is entirely back with a puff of oblivion into the utter Nothing from which we once sprang. Moreover, it might be the case that only the globally consistent laws of nature are viable and that the universe is a logically very tight structure that admits of no inconsistency or incoherence and the type of arithmetic that goes with it.

There are some related issues. Although some are pessimistic about the consequences of finding one day in the future a theory of everything, a kind of cosmic all-embracing mother not merely of all inlaws but of all laws, suggesting that the time has come for humanity to hang up its slide-rule and accept that the job has been done, with complete understanding of the ins and outs of everything, there might always be something left to do. For instance, we might find that there are two or more equally successful descriptions of everything, and not be able to choose between them. A little of that possibility has already been encountered, for as I explained in Chapter 8 it is possible to formulate a description of the world solely in terms of positions or solely in terms of momenta. Neither is the 'better' description. Maybe there are myriad seemingly irreconcilable yet equally valid descriptions of the world waiting to be discovered, myriad collections of mutually self-consistent yet seemingly disparate laws of nature.

Shall we know when we have discovered all the laws of nature? Shall we know that a particular theory of nature is valid even if its experimental verification is beyond us either technically or in principle?

With all the laws supposedly discovered, should we cautiously let slip our grip on the rigours of experimental verification or set up guards on the outposts of knowledge charged with the thankless task of identifying transgressors of our laws despite being confident that no such transgression is ever likely to occur? We shall need never-sleeping tireless always-alert robot inspectors of Nature at these frontiers of knowledge. Should we accept (as is hinted at by some contemporary fundamental theories; I have in mind string theory) that such is our confidence in our theories that even if they cannot be tested they should be accepted as true? Is our gradual prospecting for the laws of nature a fateful step towards overconfidence?

Whatever the future, it is good to know that as far as we can see, the universe is a rational place and that even the origin of the laws it abides by are within the scope of human comprehension. How I long, though, for replacing that 'not much' happening at the Creation by the breathtaking prospect that the 'not much' was nothing at all.

NOTES

The following notes refer to the *speed of light* (c), *Planck's constant* (h), *Boltzmann's constant* (k), and the *fundamental charge* (e).

Chapter 1

1. I have always admired Max Jammer's *The conceptual development of quantum mechanics* (McGraw-Hill, 1966) for its thorough treatment of the emergence of the theory.
2. Hooke's law states that $F = -k_f x$, where F is the restoring force, x is the displacement from equilibrium (the 'resting' spring), and k_f is a characteristic of the spring known as its 'force constant'. Stiff springs have big force constants. See Chapter 6 for more information.
3. One form of Boyle's law is that at constant temperature, $V \propto 1/p$, where V is the volume occupied by the gas when the pressure is p. It follows that the product pV is a constant for a given sample of gas at a constant temperature. See Chapter 6 for more information.

Chapter 2

1. An accessible account of Noether's theorem is Dwight Neuenschwander, *Emmy Noether's wonderful theorem* (Johns Hopkins University Press, 2010). For a more substantial account, try Yvette Kosmann-Schwarzbach's *The Noether theorems: invariance and conservation laws in the twentieth century*. Translated by Bertram Schwarzbach (Springer, 2011).
2. The kinetic energy of a body of mass m travelling at a speed v is $\frac{1}{2}mv^2$. The potential energy of a body of mass m at a height h above the surface of the Earth is mgh, where g is a constant, the 'acceleration of free fall' (its value is close to 9.8 m/s^2). The energy of an electromagnetic field is proportional to the squares of the strengths of its electric and magnetic fields.

3. The experimental detection of the neutrino was carried out by F. B. Harrison, H. W. Kruse, and A. D. McGuire, who were awarded the Nobel Prize for physics, but not until 1995, forty years later. Imagine being on tenterhooks for forty successive Octobers!
4. There are further worms in the can these considerations have opened. The distinction between energy and linear momentum (covered later in the chapter) depends on the state of motion of the observer and the observed, and throughout this discussion we should really be considering the uniformity of spacetime rather than each component separately. Please forgive me for ignoring that worm in the exposition (but not in my mind).
5. If the time between generations halves on each generation back, the total span of time would be finite even for an infinite number of generations ($1 + \frac{1}{2} + \frac{1}{4} + \cdots = 2$), but the time for the Ur-universe to have its daughter would have been infinitesimal. I would be disappointed if the Ur-universe emerged an infinite time ago, for that would undermine everything I say, but no doubt please those with a certain cast of mind.
6. The *Planck length* is defined as $L_P = \sqrt{hG/2\pi c^3}$, where G is the gravitational constant, and works out at about 1.6×10^{-35} metres. That's about one trillion trillionth of the diameter of an atomic nucleus. The *Planck time* is defined as the time it takes for light to travel that distance, so $t_P = \sqrt{hG/2\pi c^5}$, which works out as 5.4×10^{-44} seconds. For completeness, nothing more, I mention the *Planck mass*, $m_P = \sqrt{hc/2\pi G}$, which works out at about a reasonably imaginable 22 micrograms. A page of this book weighs about 140 000 Planck masses.
7. I am conscious of, but not distracted by, the *Hymn of Creation* in the *Rig-Veda*:

   1. Then was not non-existent nor existent: there was no realm of air, no sky beyond it.
      What covered in, and where? And what gave shelter? Was water there, unfathomed depth of water?
   2. Death was not then, nor was there aught immortal: no sign was there, the day's and night's divider. That One Thing, breathless, breathed by its own nature: apart from it was nothing whatsoever.
   3. Darkness there was: at first concealed in darkness this All was indiscriminated chaos.

8. Many (about 2×10^{52}) Planck times ago I speculated about how the emergence of something from Nothing might have taken place, in

The creation (W. H. Freeman & Co., 1981) and revisited in *Creation revisited* (W. H. Freeman & Co., 1992).

9. The linear momentum, $\boldsymbol{p}$, of a body of mass m is related to velocity, $\boldsymbol{v}$, by $\boldsymbol{p} = m\boldsymbol{v}$.
10. The angular momentum $\boldsymbol{J}$ is related to the angular velocity, $\boldsymbol{\omega}$, by $\boldsymbol{J} = I\boldsymbol{\omega}$, where I is the moment of inertia. The moment of inertia of a body of mass m rotating on a path of radius r is $I = mr^2$.

Chapter 3

1. *Snell's law of refraction* states that the angle of incidence and refraction as a ray passes through the interface of two media with refractive indices n_{r1} and n_{r2} is $\sin\theta_1 / \sin\theta_2 = n_{r2} / n_{r1}$.
2. Here is a thought if you are faced with this emergency, of someone drowning in a lake. If it is really the case that you can walk ten times faster than you can wade, and the drowning friend and you are at equal distances from the water's edge, and the same distance apart parallel to that edge, then a short but fussy calculation (which is better done now than then) shows that you should walk to a point 93 per cent of one of those distances on the edge, then wade from there.
3. Suppose the amplitude of a wave arriving at the destination by one path is a_0. The amplitude of a wave arriving by a slightly different path described by a parameter p, a measure of the bendiness of the path, is a_p. The two amplitudes are related by $a_p = a_0 + p(da/dp) + \frac{1}{2}p^2(d^2a/dp^2) + \cdots$. If the path is a minimum, the term $da/dp = 0$ and the two amplitudes differ only to second order in p; all other paths differ much more, to first order in p. Experts will know that I should be discussing the 'phase length', not the amplitude.
4. The wavelength, λ, of a particle with linear momentum of magnitude p is given by the 'de Broglie relation' $\lambda = h/p$, where h is Planck's constant (see Chapter 8), which was proposed by Louis de Broglie (1892–1987) in 1924 and later shown to be a consequence of a more general formulation of quantum mechanics.
5. The formal definition of action, S, is $S = \int_{\text{path}} L(q, \dot{q})\,ds$, where the integration is along the path with infinitesimal steps ds, q is the position of the particle, $\dot{q}$ its velocity, and $L(q, \dot{q})$ the 'lagrangian' of the system. In certain cases, L is the difference between the kinetic and potential energies of the particle, as in $L = \frac{1}{2}m\dot{q}^2 - \frac{1}{2}k_f q^2$ for a harmonic oscillator.

6. A version of quantum mechanics based on the concept of interfering paths is Feynman's 'path integral' formulation of the theory, which is set out in R. P. Feynman and A. R. Hibbs, *Quantum mechanics and path integrals* (McGraw-Hill, 1965).
7. If the wave has the amplitude a at its origin, its amplitude and phase at a distant point is $ae^{iS/\hbar}$, where S is the action associated with the path, $\hbar = h/2\pi$, and $i = \sqrt{(-1)}$.
8. Newton's second law is the differential equation $\mathbf{F} = d\mathbf{p}/dt$, where $\mathbf{F}$ is the force and $\mathbf{p}$ is the linear momentum. A more complicated example is the Schrödinger equation for a particle of mass m and energy E in a one-dimensional region where the potential energy is $V(x)$: $-(\hbar^2/2m)(d^2\psi/dx^2) + V(x)\psi = E\psi$, where $\hbar = h/2\pi$ and ψ is its 'wavefunction', a mathematical function that contains all the dynamical information about the particle.
9. To find the path corresponding to minimum action (as defined in note 5), look for the path such that $\delta \int_{\text{path}} L(q,\dot{q})ds = 0$, where δ denotes a variation in the path. This minimization is satisfied provided $\partial L/\partial q - d(\partial L/\partial \dot{q})/dt = 0$, which is a differential equation (the Euler–Lagrange equation). For a lagrangian of the form $L = \frac{1}{2}m\dot{q}^2 - V(q)$, the Euler–Lagrange equation turns into Newton's second law.

Chapter 4

1. The *Boltzmann distribution* implies that the ratio of the numbers of molecules N_1 and N_2 in states with energies E_1 and E_2 at an absolute temperature T is $N_2/N_1 = e^{-(E_2 - E_1)/kT}$. In these notes, the symbol T always means the absolute temperature.
2. Those other pedants will know that I have in mind the 'zero-point energy' of certain kinds of motion, the energy that for quantum mechanical reasons cannot be removed. It is not possible, for instance, for a pendulum to be absolutely still.
3. To obtain the absolute temperature on the Kelvin scale from the Celsius temperature, add 273.15 to the latter. So, 20 °C is 293 K.
4. The Arrhenius rate law states that the rate of a chemical reaction is proportional to $e^{-E_a/RT}$, where E_a is the energy of activation and R is the gas constant ($R = N_A k$).
5. *Newton's law of cooling* states that the difference in temperature between a body and its surroundings, ΔT, changes with time as $\Delta T(t) = \Delta T(0)e^{-Kt}$,

where K is a constant that depends on the mass and composition of the body.

6. The *law of radioactive decay* states that the number of active nuclei, N, changes with time as $N(t)=N(0)e^{-Kt}$, where K depends on the identity of the nuclide and is related to the radioactive half-life, $t_{1/2}$, by $K=(\ln 2)/t_{1/2}$.

Chapter 5

1. *Boltzmann's formula* for the entropy is $S=k \ln W$, where W is the number of ways that molecules can be distributed yet have the same total energy. The modern 'ln' denotes the natural logarithm; Boltzmann's epitaph uses log in place of ln. It would be nice to think that the writhesome letter S was chosen, by Clausius, to capture the sense of 'turning' in entropy, but I understand it was simply going spare, with its neighbours R and T already spoken for.
2. The *Clausius expression* for the change in entropy, ΔS, when energy q as heat is transferred to a body at an absolute temperature T is $\Delta S=q/T$. There are certain technical constraints on how the heat is transferred, specifically that the transfer must be carried out 'reversibly', which in practice means with a minimal temperature difference between the heated and the heater at all stages of the transfer.
3. The *efficiency*, η, of an engine is defined as the ratio of the work produced to the heat consumed. Carnot's expression for the efficiency of an ideal heat engine operating between a hot source at an absolute temperature T_{hot} and a cold sink at an absolute temperature T_{cold} is $\eta=1-T_{cold}/T_{hot}$. The efficiency approaches 1 as the temperature of the cold sink approaches zero or the temperature of the hot sink approaches infinity. High temperatures are cheaper to achieve than low temperatures, so engineers strive to raise the temperature of the hot source (superheated steam, for instance) to achieve the greatest efficiency. For a hot source at 200 °C (473 K) and a cold sink at 20 °C (293 K), the efficiency is $\eta=0.38$ (so only 38 per cent of the heat released by the fuel can be converted into work, even in an ideal engine).
4. Kelvin's words are: *It is impossible, by means of inanimate material agency, to derive mechanical effect from any portion of matter by cooling it below the temperature of the coldest of the surrounding objects.*

5. Clausius, here in translation, wrote: *Heat can never pass from a colder to a warmer body without some other change, connected therewith, occurring at the same time.*
6. For more information about the connection between the unattainability of absolute zero and the value of entropy, see my *The laws of thermodynamics: a very short introduction* (Oxford University Press, 2010), or, more rigorously, my (with Julio de Paula and James Keeler) *Physical chemistry* (11th edition, Oxford University Press, 2018).
7. To help you judge Prigogine's contribution, see I. Prigogine and I. Stengers, *The end of certainty* (The Free Press, 1997). The King of the Belgians seems to have admired, or have been advised to admire, his work, for Prigogine was created viscount in 1989.

Chapter 6

1. The *perfect gas law* is $pV = NkT$. Where p is the pressure, V the volume, N the number of molecules present, T the absolute temperature. Chemists typically write the law in terms of the amount of molecules present, n, with $n = \mathrm{N}/\mathrm{N}_\mathrm{A}$, N_A being *Avogadro's constant*, and write $\mathrm{N}_\mathrm{A}k = R$, the *gas constant*. It then has the form $pV = nRT$.
2. Most people refer to the law as the 'ideal gas law'. However, I like to stick to 'perfect'. The reason is as follows. There are things called 'ideal solutions' where molecules of solute and solvent interact with each other but one molecule doesn't know whether its neighbour is a solute or a solvent molecule: the interactions between them are the same. That is true in a perfect gas, but not only are the interactions between molecules the same, they are also zero. So perfection is one further step along the road than ideality.
3. *Henry's law* states that the concentration, c, of gas in a liquid at equilibrium is proportional to the pressure of the gas ($c = Kp$); *Raoult's law* states that the presence of a solute lowers the vapour pressure of the solvent by Δp in proportion to the concentration of the solute ($\Delta p = Kc$); *van 't Hoff's law* states that the osmotic pressure, Π, is proportional to the concentration of the solute ($\Pi = Kc$). The Ks are different in each case.
4. The mathematical treatment results in the expression $pV = \frac{1}{3}Nmv_{\mathrm{rms}}^2$, where N is the number of molecules in the volume V, m is the mass of a molecule, and v_{rms} is the root-mean-square speed, the square root of the average value of the squares of the speeds of the molecules. Think of it

loosely as their average speed. At constant temperature this expression has the form $pV = \text{constant}$, which is Boyle's law.

5. The average (mean) speed of molecules of mass m in a gas at an absolute temperature T is $v_{\text{mean}} = (8kT / \pi m)^{1/2}$. That is, $v_{\text{mean}} \propto \sqrt{T}$.
6. As remarked in the first note to Chapter 1, *Hooke's law* states that $F = -k_f x$, where F is the restoring force and x is the displacement from equilibrium. The frequency of an oscillator of mass m is $\nu = (1/2\pi)(k_f/m)^{1/2}$. For a pendulum of length l, $\nu = (1/2\pi)(g/l)^{1/2}$, where g is the 'acceleration of free fall', a measure of gravitational pull. The latter result is also limiting in the sense that it is exact only in the limit of the swing being zero.
7. The most general expression for a property that has the value $P(x)$ when the displacement from equilibrium is x is $P(x) = P(0) + (\mathrm{d}P/\mathrm{d}x)_0 x + \frac{1}{2}(\mathrm{d}^2P/\mathrm{d}x^2)_0 x^2 + \cdots$. At the minimum of the curve showing how P depends on x, $(\mathrm{d}P/\mathrm{d}x)_0 = 0$. Therefore, the first non-vanishing term after $P(0)$ is $\frac{1}{2}(\mathrm{d}^2P/\mathrm{d}x^2)_0 x^2$. If P is the potential energy E_p, then because the restoring force F and the potential energy are related by $F = -\mathrm{d}E_p/\mathrm{d}x$, in this general scenario, $F = -(\mathrm{d}^2P/\mathrm{d}x^2)_0 x$, which is Hooke's law when $(\mathrm{d}^2P/\mathrm{d}x^2)_0$ is identified with k_f.
8. The energy of a harmonic oscillator, an oscillator that obeys Hooke's law, is $E = p^2/2m + (k_f/2)x^2$, where m is its mass. Note the symmetry: both the linear momentum p and the displacement x occur as their squares.
9. The structure and its diffraction pattern are essentially each other's Fourier transform. A 'position description' and a 'momentum description' of the world are likewise each other's Fourier transform.

Chapter 7

1. *Coulomb's inverse square law* identifies the magnitude of the force between two electric charges Q_1 and Q_2 separated by a distance r as $F = Q_1Q_2 / 4\pi\varepsilon_0 r^2$, where ε_0 is a fundamental constant, the *vacuum permittivity*. The potential energy of the two charges is $E_p = Q_1Q_2 / 4\pi\varepsilon_0 r$. A similar inverse-square law expresses the magnitude of the gravitational force between two masses m_1 and m_2 as $F = Gm_1m_2 / r^2$, where G is the *gravitational constant*.
2. The full group-theoretical designation of the symmetry of the Coulomb interaction is SO(4), the 'special orthogonal group in four dimensions'.
3. In a hydrogen atom, atomic orbitals of the same shell (as designated by the principal quantum number n) all have the same energy regardless

of their angular momentum around the nucleus (as designated by the angular momentum quantum number l). Thus, s, p, d... orbitals of the same shell all have the same energy. 'Degeneracy', the possession of the same energy, is always associated with symmetry; in this case it is a consequence of the four-dimensional hypersphericity of the Coulomb interaction, which allows these orbitals with their various shapes to be rotated into one another in four dimensions.

4. If the original wave is $\psi(x)$, under a global gauge transformation, a uniform shift of phase through the angle ϕ, it becomes $\psi(x)\mathrm{e}^{\mathrm{i}\phi}$. The probability density of the particle is $\psi^*(x)\,\psi(x)$ before the transformation and becomes $\psi^*(x)\,\mathrm{e}^{-\mathrm{i}\phi}\psi(x)\,\mathrm{e}^{\mathrm{i}\phi}=\psi^*(x)\psi(x)$ after the transformation. It is invariant. This invariance survives under a local gauge transformation $\phi(x)$ because still $\psi^*(x)\,\mathrm{e}^{-\mathrm{i}\phi(x)}\psi(x)\,\mathrm{e}^{\mathrm{i}\phi(x)}=\psi^*(x)\psi(x)$.
5. Here is the technical argument relating a global gauge transformation to the conservation of charge. I am stripping away as much of the notation as possible and aim simply to show the pathway through the argument: to do it properly you would need to consider time derivatives as well as the single space derivative used here. Consider an infinitesimal nudge, such that the transformation $\psi(x)\rightarrow \mathrm{e}^{\mathrm{i}\phi}\psi(x)$ can be approximated by $\psi(x)\rightarrow(1+\mathrm{i}\phi)\psi(x)=\psi(x)+\delta\psi(x)$ with $\delta\psi(x)=\mathrm{i}\phi\,\psi(x)$. The resulting change in the lagrangian density $L(\psi,\psi')=\tfrac{1}{2}\psi'^2-\tfrac{1}{2}m\psi^2$, where $\psi'=\partial\psi/\partial x$, is

$$\delta L=\frac{\partial L}{\partial\psi}\delta\psi+\frac{\partial L}{\partial\psi'}\delta\psi'=\left\{\frac{\partial L}{\partial\psi}-\frac{\partial}{\partial x}\left(\frac{\partial L}{\partial\psi'}\right)\right\}\delta\psi+\frac{\partial}{\partial x}\left(\frac{\partial L}{\partial\psi'}\delta\psi\right)$$

Note that, according to the Euler–Lagrange equation (the equation that tells you how to grope along the path so as to minimize the action overall),

$$\frac{\partial L}{\partial\psi}-\frac{\partial}{\partial x}\left(\frac{\partial L}{\partial\psi'}\right)=0$$

Therefore

$$\delta L=\frac{\partial}{\partial x}\left(\frac{\partial L}{\partial\psi'}\delta\psi\right)=\mathrm{i}\phi\frac{\partial}{\partial x}\psi'\psi$$

The lagrangian density is unchanged under the global gauge transformation, so $\delta L=0$ for arbitrary ϕ. Therefore,

$$\frac{\partial}{\partial x}\overbrace{\psi'\psi}^{J}=0$$

and the current J is conserved.

6. Suppose the wavefunction $\psi(x)$ satisfies the Schrödinger equation

$$-\frac{\hbar^2}{2m}\frac{\mathrm{d}^2\psi(x)}{\mathrm{d}x^2}+V(x)\psi(x)=E\,\psi(x)$$

Now shift the phase of the wavefunction to $\psi(x)\mathrm{e}^{\mathrm{i}\phi(x)}=\tilde{\psi}(x)$. This phase-shifted function no longer satisfies the same equation, because

$$\begin{aligned}-\frac{\hbar^2}{2m}\frac{\mathrm{d}^2\tilde{\psi}(x)}{\mathrm{d}x^2}&+V(x)\,\tilde{\psi}(x)\\ &=-\frac{\hbar^2}{2m}\left\{\frac{\mathrm{d}^2\psi}{\mathrm{d}x^2}+2\mathrm{i}\frac{\mathrm{d}\phi}{\mathrm{d}x}\frac{\mathrm{d}\psi}{\mathrm{d}x}-\left(\frac{\mathrm{d}\phi}{\mathrm{d}x}\right)^2\psi+\mathrm{i}\frac{\mathrm{d}^2\phi}{\mathrm{d}x^2}\psi\right\}\mathrm{e}^{\mathrm{i}\phi(x)}\\ &\quad+V(x)\,\tilde{\psi}(x)\\ &=E\,\tilde{\psi}(x)-\frac{\hbar^2}{2m}\left\{2\mathrm{i}\frac{\mathrm{d}\phi}{\mathrm{d}x}\frac{\mathrm{d}\psi}{\mathrm{d}x}-\left(\frac{\mathrm{d}\phi}{\mathrm{d}x}\right)^2\psi+\mathrm{i}\frac{\mathrm{d}^2\phi}{\mathrm{d}x^2}\psi\right\}\mathrm{e}^{\mathrm{i}\phi(x)}\end{aligned}$$

The unwanted three additional terms are eliminated if the Schrödinger equation is modified to

$$-\frac{\hbar^2}{2m}\frac{\mathrm{d}^2\tilde{\psi}(x)}{\mathrm{d}x^2}+U(x)\tilde{\psi}(x)+V(x)\tilde{\psi}(x)=E\tilde{\psi}(x)$$

with

$$U(x)=\frac{\hbar^2}{2m}\left\{2\mathrm{i}\left(\frac{\mathrm{d}\phi}{\mathrm{d}x}\right)\frac{\mathrm{d}}{\mathrm{d}x}+\left(\frac{\mathrm{d}\phi}{\mathrm{d}x}\right)^2+\mathrm{i}\frac{\mathrm{d}^2\phi}{\mathrm{d}x^2}\right\}$$

The additional tem, $U(x)$, is like the energy contribution $V(x)$ and is an interaction term with the field. Thus, interactions emerge from local gauge invariance. Note that the term proportional to d/dx is in fact proportional to the linear momentum operator, $p=(\hbar/\mathrm{i})\mathrm{d}/\mathrm{d}x$.

Chapter 8

1. The relation between frequency, ν, and the energy, E, of the associated quanta, is $E = h\nu$ and in conventional units $h = 6.626 \times 10^{-34}$ joule-seconds. It follows that an energy (in joules) divided by Planck's constant is a frequency (a quantity 'per second'). An energy of 1 joule converts in this way to about 2×10^{33} cycles per second. Planck's relation becomes $E^{\dagger} = \nu$, and his constant has vanished. If you insist on preserving the form $E^{\dagger} = h\nu$, you can do so, but you have to set $h = 1$.
2. The Schrödinger equation for a particle of mass m moving in a region where its potential energy is V and its total energy is E is

$$-\frac{h^2}{8\pi^2 m}\nabla^2\psi + V\psi = E\psi$$

With $mc^2/h = m^{\dagger}$, $V/h = V^{\dagger}$, and $E/h = E^{\dagger}$, $c\nabla = \nabla^{\dagger}$

$$-\frac{1}{8\pi^2 m^{\dagger}}\nabla^{\dagger 2}\psi + V^{\dagger}\psi = E^{\dagger}\psi$$

which looks the same except for the vanishing of h. A standard problem in elementary quantum mechanics is to find the allowed energy levels of a particle confined to a region of space of length L. The conventional solution is $E = n^2h^2/8mL^2$ with $n = 1,2,\ldots$. With $h = 1$, the solutions are $E^{\dagger} = n^2/8m^{\dagger}L^{\dagger 2}$ where $L^{\dagger} = L/c$. Another basic solution is that of a harmonic oscillator, which is conventionally written $E = (n+\frac{1}{2})h\nu$ with $n = 0,1,2,\ldots$, $\nu = (1/2\pi)\sqrt{k_f/m}$, and $k_f = (d^2V/dx^2)_0$. With $h = 1$ the solutions are $E^{\dagger} = (n+\frac{1}{2})\nu$ with $\nu = (1/2\pi)\sqrt{k_f^{\dagger}/m^{\dagger}}$ and $k_f^{\dagger} = (d^2V^{\dagger}/dx^{\dagger 2})_0$, with $x^{\dagger} = x/c$, where I have used a notation that will be meaningful to those who are familiar with these matters.
3. If position is denoted x and linear momentum along the same direction is denoted p, then the commutation of position and linear momentum is $xp - px$. This combination is normally denoted by $[x,p]$, which is the 'commutator' of x and p. In quantum mechanics, x and p are treated as 'operators' (things that do things, such as multiplying or differentiating a function) and its entire edifice springs from the relation $[x,p] = ih/2\pi$, where i is the 'imaginary number' $\sqrt{(-1)}$. You could take the view that the foundations of quantum mechanics are entirely imaginary.
4. With $x^{\dagger} = x/c$ and $m^{\dagger} = mc^2/h$, linear momentum becomes $p^{\dagger} = cp/h$. The commutator $[x,p] = ih/2\pi$ then becomes $[x^{\dagger},p^{\dagger}] = i/2\pi$. Suppose that at some instant you, of mass 70 kilograms, are at a location we can call

2 metres from a certain point and are travelling at 3 metres per second. Your linear momentum (the product of mass and velocity) is 70 kg × 3 m/s = 210 kg m/s. The product of your position and momentum is 2 m × 210 kg m/s = 420 kg m^2/s. Planck's constant, in the same units, is 6.6×10^{-34} kg m^2/s, vastly smaller, and utterly negligible. With the new system of units, 2 m from somewhere is actually 7 nanoseconds from there, and a mass of 70 kg moving at 3 m/s is actually a momentum, expressed as a frequency, of 9.5×10^{43} Hz. So the product of position and momentum is 7×10^{35}, which is far, far greater than 1.

5. One consequence of the discarding of k is that entropy (remember how Boltzmann defined it as $S = k \log W$) becomes $S = \log W$, and has become a pure number without units. The perfect gas law $pV = NkT$ becomes $pV = NT^{\dagger}$ If you insist on preserving the conventional form of the gas law and writing it as $pV = NkT^{\dagger}$, then you have to accept that $k = 1$. The gas law is commonly written $pV = nRT$, where n is the amount of molecules (in moles) and R is the gas constant. The latter is related to k by $R = N_A k$, where N_A is Avogadro's constant. With $k = 1$, $R = N_A$.
6. Here, side by side, are four examples to show how the use of $\mathcal{T}$ simplifies the appearance of equations:

	Conventional	Revised
Perfect gas law	$pV = NkT$	$pV\mathcal{T} = N$
Boltzmann distribution	$N_2/N_1 = e^{-(E_2-E_1)/kT}$	$N_2/N_1 = e^{-\mathcal{T}(E_2-E_1)}$
Energy of N harmonic oscillators	$E = \dfrac{Nh\nu}{e^{h\nu/kT} - 1}$	$E = \dfrac{Nh\nu}{e^{\mathcal{T}h\nu} - 1}$
...and their heat capacity	$C = Nk\left(\dfrac{h\nu}{kT}\right)^2 \left(\dfrac{e^{-h\nu/2kT}}{1-e^{-h\nu/kT}}\right)^2$	$C = Nk\left(\dfrac{\mathcal{T}h\nu e^{-\mathcal{T}h\nu/2}}{1-e^{-\mathcal{T}h\nu}}\right)^2$

7. Here I am introducing $\mathcal{T}^{\dagger} = h\mathcal{T} = h/kT$.
8. Here are the final forms of the four quantities in note 6 where, in addition to the quantities already introduced, $C^{\dagger} = C/k$, $V^{\dagger} = V/c^3$, and $p^{\dagger} = c^3p/h$, which are respectively dimensionless, in units of seconds^3, and in units of 1/seconds^4.

Perfect gas law	$p^{\dagger}V^{\dagger}\mathcal{T}^{\dagger} = N$
Boltzmann distribution	$N_2 / N_1 = e^{-\mathcal{T}^{\dagger}(E_2{}^{\dagger} - E_1{}^{\dagger})}$
Energy of N harmonic oscillators	$E^{\dagger} = \dfrac{N\nu}{e^{\mathcal{T}^{\dagger}\nu} - 1}$
...and their heat capacity	$C^{\dagger} = N\left(\dfrac{\mathcal{T}^{\dagger}\nu e^{-\mathcal{T}^{\dagger}\nu/2}}{1 - e^{-\mathcal{T}^{\dagger}\nu}}\right)^2$

9. The fine-structure constant is defined as $\alpha = \mu_0\varepsilon^2\chi / 2\eta$, where $\propto_0$ is the vacuum permeability, with the defined value of $4\pi \times 10^{-7}$ J s² C⁻² m⁻¹. A more precise value than $\alpha = 1/137$ is $\alpha = 0.007\,297\,352\,5664$. There is some arbitrariness in the definition of α because there might be a more fundamental measure of electric charge. The charge of a quark, for instance, is $\frac{1}{3}e$, and that might be a more appropriate value. In that case, α would work out to be nine times smaller, at around 1/1233.
10. One such concoction that purports to reproduce the value of the fine-structure constant is $\alpha = 29\cos(\pi/137)\tan(\pi/(137\times29))/\pi$, which works out to 0.007 297 352 531 86....
11. I referred to the inverse square law in note 1 of Chapter 6: it expressed the magnitude of the gravitational force between two masses m_1 and m_2 as $F = Gm_1m_2 / r^2$. The gravitational constant G has the value 6.673×10^{-13} kg⁻¹ m³ s⁻².
12. The dimensionless form of G is $\alpha_G = 2\pi G m_e^2 / hc$. There is no fundamental reason for selecting the electron mass in this definition, only analogy with the e^2 that appears in α, so the numerical value of α_G is perhaps only suggestive of the strength of gravity.

Chapter 9

1. E. P. Wigner (1960). *The unreasonable effectiveness of mathematics in the natural sciences.* Richard Courant lecture in mathematical sciences delivered at New York University, 11 May 1959. *Communications on Pure and Applied Mathematics.* 13: 1–14.
2. The following discussion of the mathematical underpinning of reality draws on ideas I published in *Creation* (1983) and *Creation revisited* (1992). Some decades later, in his book *Our mathematical universe* (2014), Max Tegmark, perhaps independently, advanced similar ideas.

3. For an introduction to the equations underlying patterns on animal pelts, see chapter 15 of *Mathematical biology*, J. D. Murray (Springer Verlag, 1989).
4. The Peano axioms are (in an abbreviated form) as follows:

    1. 0 is a natural number.
    2. For every natural number n, its successor is a natural number.
    3. For all natural numbers m and n, $m = n$ if and only if the successors of m and n are equal.
    4. There is no natural number whose successor is 0.

    Addition (+) is then defined such that $n + 0 = n$ and $n + S(m) = S(n + m)$, and *multiplication* (×) is defined such that $n \times 0 = 0$ and $n \times S(m) = n + (n \times m)$, where $S(n)$ is the successor of n.
5. In its early, but still not particularly accessible, form the Löwenheim–Skolem theorem states that states that *if a countable first-order theory has an infinite model, then for every infinite cardinal number* κ *it has a model of size* κ. A more digestible consequence is that a system of rules like those of arithmetic emulates any field of knowledge that can be formalized as a set of axioms.
6. Specifically, for all fermions (particles with half-integral spin, which includes electrons), under interchange of the labels of two identical fermions the wavefunction must change sign: $\psi(2,1) = -\psi(1,2)$. The principle is deeply rooted in relativity.
7. The Wikipedia article is very clear: '[Gödel] proved for any computable axiomatic system that is powerful enough to describe the arithmetic of the natural numbers (e.g. the Peano axioms or Zermelo–Fraenkel set theory with the axiom of choice), that:

    > If a (logical or axiomatic formal) system is consistent, it cannot be complete.
    > The consistency of the axioms cannot be proved within the system.

    These theorems ended a half-century of attempts, beginning with the work of Frege and culminating in *Principia Mathematica* and Hilbert's formalism, to find a set of axioms sufficient for all mathematics.' An English version of 'On formally undecidable propositions' can be found in *Gödel's theorem in focus*, ed. S. G. Shanker (Routledge, 1988), but you will have to grapple with statements like '$0 \text{ St } v, x = \varepsilon n \mid n \leq l(x) \ \&\ \text{Fr } n, x \ \&\ \overline{(Ep)}[n < p \ldots$'. A much more accessible version for Earthbound humans is *Gödel's proof*, E. Nagel and J. R. Newman (Routledge, 1958).
8. Here I have in mind what is called 'Presburger arithmetic', which is Peano *sans* ×. For a very nice account, accessible too, see John Barrow's *New theories of everything* (Oxford University Press, 2007).

INDEX

THE OXFORD BOOK OF MODERN SCIENCE WRITING

Edited by Richard Dawkins

978-0-19-921681-9 | Paperback | £10.99

'A compendium of some of the most illuminating thinking of the past 100 years.'

The Times

'A feast for many long evenings.'

The Sunday Telegraph

Selected and introduced by Richard Dawkins, The Oxford Book of Modern Science Writing is a rich and vibrant anthology celebrating the finest writing by scientists for a wider audience—revealing that the best scientists have displayed as much imagination and skill with the pen as they have in the laboratory.

ELEGANCE IN SCIENCE

The Beauty of Simplicity

Ian Glynn

978-0-19-966881-6 | Paperback | £10.99

'An erudite book... Well illustrated and full of historical anecdote and background, this is an elegant volume indeed.' ***Nature***

'There is a wealth of historical information packed in here.' ***Times Literary Supplement***

The idea of elegance in science is not necessarily a familiar one, but it is an important one. The use of the term is perhaps most clear-cut in mathematics—the elegant proof—and this is where Ian Glynn begins his exploration. Scientists often share a sense of admiration and excitement on hearing of an elegant solution to a problem, an elegant theory, or an elegant experiment. With a highly readable selection of inspiring episodes highlighting the role of beauty and simplicity in the sciences, this book also relates to important philosophical issues of inference, and Glynn ends by warning us not to rely on beauty and simplicity alone: even the most elegant explanation can be wrong.

WHAT IS CHEMISTRY?

Peter Atkins

978-0-19-968398-7 | Hardback |£11.99

'Atkins wins his readers' attention simply through an elegant and lucid description of the subject he loves.' ***Nature***

In *What is Chemistry?* Peter Atkins encourages us to look at chemistry anew, through a chemist's eyes, to understand its central concepts and to see how it contributes not only towards our material comfort, but also to human culture. He shows how chemistry provides the infrastructure of our world, through the chemical industry, the fuels of heating, power generation, and transport, as well as the fabrics of our clothing and furnishings. By considering the remarkable achievements that chemistry has made, and examining its place between both physics and biology, Atkins presents a fascinating, clear, and rigorous exploration of the world of chemistry.

WHAT IS LIFE?

How Chemistry Becomes Biology

Addy Pross

978-0-19-878479-1 | Paperback |£9.99

'Pross does an excellent job of succinctly conveying the difficulty in crafting an unambiguous general definition of life and provides a roadmap to much of the work on the origin of life done by chemists in the past 50 years. The book is worth the read for these discussions alone.' ***Chemical Heritage***

Living things are hugely complex and have unique properties, such as self-maintenance and apparently purposeful behaviour which we do not see in inert matter. So how does chemistry give rise to biology? What could have led the first replicating molecules up such a path? Now, developments in the emerging field of 'systems chemistry' are unlocking the problem. The gulf between biology and the physical sciences is finally becoming bridged.

REACTIONS

The Private Life of Atoms

Peter Atkins

978-0-19-966880-9 | Paperback | £12.99

'The perfect antidote to science phobia.'

Booklist

Peter Atkins captures the heart of chemistry in this book, through an innovative, closely integrated design of images and text, and his characteristically clear, precise, and economical exposition. Explaining the processes involved in chemical reactions, he begins by introducing a 'tool kit' of basic reactions, such as precipitation, corrosion, and catalysis, and concludes by showing how these building blocks are brought together in more complex processes such as photosynthesis, to provide a concise and intellectually rewarding introduction to the private life of atoms.

HYPERSPACE

A Scientific Odyssey through Parallel Universes, Time Warps, and the Tenth Dimension

Michio Kaku

978-0-19878503-3 | Paperback | £9.99

'Beautifully written, making difficult scientific ideas seem accessible, almost easy.'

Danah Zohar, The Independent

'Since ingesting Einstein's relativity theory, physics fell down a quantum rabbit hole and, ever since, physicists' reports to the world of popular science have been curiouser and curiouser... [Kaku] delineates the 'delicious contradictions' of the quantum revolution. His intellectual perceptions will thrill lay readers, SF fans and the physics-literate.' ***Publisher's Weekly***

Michio Kaku's classic book describes the development of ideas about multidimensional space. He takes the reader on a ride through hyperspace to the edge of physics. On the way he gives crystal clear explanations of such formidable mathematical concepts as non-Euclidean geometry, Kaluza-Klein Theory, and Supergravity, the everyday tools of the string theorist. Utilizing fascinating and often hilarious anecdotes from history, art, and science fiction, he shows us that writers and artists—in addition to scientists—have been fascinated by multidimensional space for over a century.

ON BEING

A scientist's exploration of the great questions of existence

Peter Atkins

978-0-19-966054-4 | Paperback | £8.99

'Crisp with good sense, clear with scientific knowledge effortlessly imparted, and delicious with the sort of wit that makes you stop and put the book down just to enjoy it the more fully. It presents a vision of life and death, of matter and space and time that is honest and consistent and miracle-free.'

Philip Pullman

'Few can match the chiseled beauty of Peter Atkins's prose as he reflects on the nature of life and death, of beginnings and endings.'

Richard Dawkins

'A paean to science.' ***Times Literary Supplement***

In this scientific 'Credo', Peter Atkins considers the universal questions of origins, endings, birth, and death to which religions have claimed answers. With his usual economy, wit, and elegance, unswerving before awkward realities, Atkins presents what science has to say. While acknowledging the comfort some find in belief, he declares his own faith in science's capacity to reveal the deepest truths.

ORIGINS

The Scientific Story of Creation

Jim Baggott

978-0-19-870765-3 | Paperback | £16.99

'*Origins* recounts the greatest story ever told: the evolution of the Universe since the Big Bang. This rich cross-disciplinary tale reminds us that astronomy, physics, chemistry, geoscience, biology and neuro science are interconnected. Baggott takes the reader on a linear, 13.8-billion-year journey. He reminds us that big questions remain in this most wonderful scientific adventure.' ***Nature***

'An impressive scientific *tour-de-force*.'
Chemistry & Industry

What is life? Where do we come from and how did we evolve? What is the universe and how was it formed? What is the nature of the material world? How does it work? How and why do we think? What does it mean to be human? How do we know?

There are many different versions of our creation story. This book tells the version according to modern science, starting at the Big Bang and travelling right up to the emergence of humans as conscious intelligent beings, 13.8 billion years later. Chapter by chapter, it sets out the current state of scientific knowledge: the origins of space and time; energy, mass, and light; galaxies, stars, and our sun; the habitable earth, and complex life itself. Baggott recounts what we currently know of our history, highlighting the questions science has yet to answer.